SpringerBriefs in Space Development

Guest Editors:
Martin Sweeting, Yaw Nyampong, Peter Marshall

For further volumes:
http://www.springer.com/series/10058

Ram S. Jakhu • Joseph N. Pelton

Small Satellites
and Their Regulation

Ram S. Jakhu
McGill University Faculty Law
Montreal, QC, Canada

Joseph N. Pelton
Arlington, USA

ISSN 2191-8171   ISSN 2191-818X (electronic)
ISBN 978-1-4614-9422-5   ISBN 978-1-4614-9423-2 (eBook)
DOI 10.1007/978-1-4614-9423-2
Springer New York Heidelberg Dordrecht London

Library of Congress Control Number: 2013952920

Printed on acid-free paper

Springer is part of Springer Science+Business Media (www.springer.com)

*This book is dedicated to all the people
who are seeking better answers to the
increasing number of issues arising
from the exploration and use of outer
space. This includes the staff of the UN
Office of Outer Space Affairs, the members
of the UN Committee on the Peaceful Uses
of Outer Space, the Inter-Agency Space
Debris Coordinating (IADC) Committee, and
others from relevant international
and national agencies around the world.*

This Springer book is published in collaboration with the International Space University (ISU). At its central campus in Strasbourg, France, and at various locations around the world, the ISU provides graduate-level training to the future leaders of the global space community. The university offers a 2-month Space Studies Program, a 5-week Southern Hemisphere Program, a 1-year Executive MBA, and a 1-year Master's Program related to space science, space engineering, systems engineering, space policy and law, business and management, and space and society.

These programs give international graduate students and young space professionals the opportunity to learn while solving complex problems in an intercultural environment. Since its founding in 1987, the International Space University has graduated more than 3,000 students from 100 countries, creating an international network of professionals and leaders. ISU faculty and lecturers from around the world have published hundreds of books and articles on space exploration, applications, science, and development.

Acknowledgments

We wish to acknowledge with sincere gratitude the invaluable assistance provided to us by Dr. David Finkleman (a leading authority on civil, commercial, and military space systems). Dr. Finkleman authorized us to use in this book some parts of his very short but excellent unpublished paper entitled "The Small Satellite Dilemma." We are thankful to both Sir Martin Sweeting and Dr. Yaw Nyampong for reviewing and providing highly valuable comments on earlier drafts of this book that improved the quality and accuracy of this book.

We also wish to express our thanks to Dr. Setsuko Aoki, Dr. Tare Brisibe, Ms. Irene Ekweozoh, Dr. Thomas Gillon, Dr. Atsuyo Ito, Ms. Joy-Marie Lawrence, Ms. Justine Limpitlaw, Mr. Bruce Mann, and Mr. K. R. Sridhara Murthi for supplying very useful information that also was used to improve the contents and factual accuracy of some parts of this book. In addition, we would like to express our appreciation to Dr. Delbert Smith and Elisabeth Evans of Jones Day Law firm; Alistair Scott, President of the British Interplanetary Society; Sir Martin Sweeting of the Surrey Space Centre and Surrey Space Technology, Ltd.; and Mr. Peter Marshall for their assistance in this undertaking in various ways. As always and notwithstanding the invaluable contributions mentioned above, we remain exclusively responsible for any errors contained in this book.

About the Authors

Ram S. Jakhu

Dr. Ram S. Jakhu is Associate Professor at the Institute of Air and Space Law, Faculty of Law, McGill University, Montreal, Canada, where he teaches and conducts research in international space law, law of space applications, law of space commercialization, government regulation of space activities, law of telecommunications and Canadian communications law, and public international law. He is a member of the Space Security Council of the World Economic Forum and a Fellow as well as the Chairman of the Legal and Regulatory Committee of the International Association for the Advancement of Space Safety (IAASS). In 2007, he received a "Distinguished Service Award" from the International Institute of Space Law for significant contributions to the development of space law. He is managing editor of the Space Regulations Library series and a member of the editorial boards of the *Annals of Air and Space Law* and of the *German Journal of Air and Space Law*. He served as a member of the Board of Directors of the International Institute of Space Law, 1999–2013; as Director, Centre for the Study of Regulated Industries, McGill University, 1999–2004; and as the First Director of the Master's Program of the International Space University, Strasbourg, France, 1995–1998. He is a widely published author and editor of an award-winning book, *National Regulation of Space Activities*. His academic degrees include a B.A. as well as an LL.B. from Panjab University, an LL.M. also from Panjab University in International Law, an LL.M. from McGill University in Air and Space Law, and a Doctor of Civil Law (on Dean's Honors List) from McGill University in Law of Outer Space and Telecommunications.

Joseph N. Pelton

Joseph N. Pelton, Ph.D., is Principal of Pelton Consulting International. He is the former President of the International Space Safety Foundation and a member of the Executive Board of the International Association for the Advancement of Space Safety. He is the former Chairman of the Board of Trustees and Vice President and Dean of the International Space University as well as the Director Emeritus of the Space and Advanced Communications Research Institute (SACRI) at George Washington University. Dr. Pelton also served as Director of the Accelerated Master's of Science Program in Telecommunications and Computers at George Washington University from 1998 to 2005. He was the founder of the Arthur C. Clarke Foundation and remains as the Vice Chairman on its Board of Directors.

Pelton is a widely published and award-winning author with over 30 books written or coauthored or coedited with colleagues. His book *Global Talk* was nominated for a Pulitzer and won the Eugene M. Emme Astronautical Literature Award. Dr. Pelton is a full member of the International Academy of Astronautics, an Associate Fellow of the American Institute of Aeronautics and Astronautics (AIAA), and a Fellow of the International Association for the Advancement of Space Safety (IAASS). He was the Founding President of the Society of Satellite Professional International (SSPI) and a member of the SSPI Hall of Fame. For the last 2 years, he has served as President of the Comsat Alumni and Retirees Association (COMARA). He received his degrees as follows: a B.S. from the University of Tulsa, an M.S. from New York University, and his doctorate from Georgetown University.

Contents

Chapter 1
Why Small Satellites and Why This Book?

1.1 Introduction

At the beginning of the Space Age, all satellites tended to be small due to limited lift capabilities of early launch vehicles. The weight (mass) of early satellites such as Explorer I and Intelsat I (or Early Bird) only ranged within tens of kilograms. As launchers and rocket systems became more capable and experimental satellites were designed to carry out more sophisticated missions, satellites became bigger and more massive due to economies of scale and increasing global demand. Space stations designed to support humans aboard have become massive. Nevertheless, a number of space applications continue to make sense for small satellites (i.e., cube-sats, micro satellites, nano satellites, or small spacecraft within a constellation). Such small satellites still make sense for a number of different financial, operational, or technical reasons. In some cases, constellations of small satellites can accomplish feats that one large satellite cannot.

Currently, there is an ongoing revolution in the development and deployment of small satellites. At present, there are more than 50 cube-sats in long-lived orbits.[1] Approximately 90 % of small satellites reside and operate in low Earth orbit and hundreds more are slated to be launched to this type of orbit in the not-too-distant future. An increasing number will join the quite crowded Sun-synchronous polar orbit.[2] NASA recently solicited concepts for debris mitigation from cube-sats. Clearly, the world is taking these small spacecraft seriously and the hazards that orbital debris is now posing to space safety. Orbital debris of all types, including small satellites, can have a potentially unfavorable impact on all types of future space enterprises.

---

[1] D.L. Oltrogge, and K. Leveque, An Evaluation of Cube-sat Orbital Decay, SSC11-VII-2, AIAA/Utah State University, Small Satellite Conference, August 2011.

[2] Giovanni Verlini, The Bright Future of Small Satellite Technology, Via *Satellite*, August 1, 2011.

R.S. Jakhu and J.N. Pelton, *Small Satellites and Their Regulation*, SpringerBriefs in Space Development, DOI 10.1007/978-1-4614-9423-2_1, © Springer New York 2014

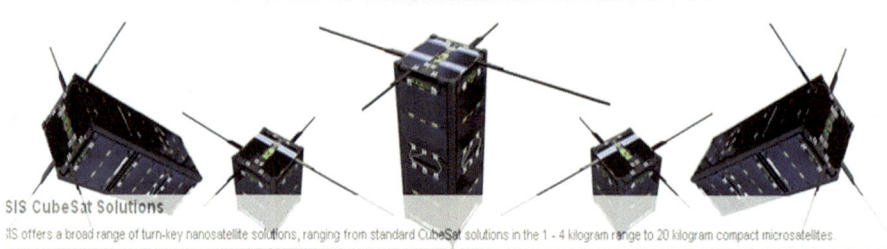

Fig. 1.1 Typical cube-sat configurations offered by a number of vendors (Courtesy of ISIS Cube-sat Solutions)

Each nation has a different perspective on small satellites. Developing nations are enthusiastic, and the U. N. Office of Outer Space Affairs is promoting that interest with conferences and workshops around the world. Undoubtedly, these satellites will be designed and built with diverse instruments by private companies and universities all over the world. Figure 1.1 demonstrates various cube-sat configurations that are commercially available and affordable.

Often "small" satellites are thought of as if they are all the same. The definition of smallness is controversial. There are several different definitions of small satellites. The most common definition seems to be the one presented in the IAA study of Earth observation satellites:

- Mini satellites are less than 1,000 kg
- Micro satellites are less than 100 kg
- Nano satellites are less than 10 kg
- Pico satellites are less than 1 kg

However, these mass-based categories alone are not sufficient to characterize satellite orbital risks or architectures. Mass alone does not define shape, orientation, maneuverability, observability, and other important features. The choice of orbital architectures for small satellites by any definition must also consider these other characteristics. In short, physical size, shape, and mass are important, but other characteristics can be key for mitigating debris and moderating orbit traffic. These include the ability to observe the spacecraft for the purpose of orbit determination, the ability of the spacecraft to maneuver, and the ability to communicate with and control the satellite.

In fact there are many different types of small satellites, with multiple functions and users. The following chart (Table 1.1) provides a useful way to sort out the many different types of small satellites based on different functions and sizes.

The above chart is only a sampling of the various functions and types of small satellites by some useful categories. The reality is, of course, much more complex. Constellations of small satellites that are relatively few in number, such as the 2-satellite Lifesat, might be used as store and forward services for public or humanitarian services and have connection times that might be measured in hours. A larger group such as the Orbcomm constellation (with nearly 30 small satellites) might be

Table 1.1 Different functional types and sizes of small satellites

Functional types and sizes of small satellites

Sizes (by mass)	Telecom constellation	Message data relay	Amateur radio	Remote sensing low res active sensing	Systems receiving signals from ground or sea-based sensors	Meteorological	Scientific experiments	Student & university experiments
Small 100–1,500 kg	Typical	Typical		Typical	Typical	Typical	Typical	Occasional
Micro 10–99 kg	Occasion-al	Typical	Typical	Occasional	Typical	Typical	Typical	Typical
3U Cube-sat 10–20 kg	Typical	Typical	Typical	Occasional	Typical	Occasional	Typical	Typical
Cube 5–10 kg	Rare	Rare	Rare				Occasional	Typical
Nano, pico or femto sats[a]							Rare	Typical

[a]Definitions can vary, but a nano-satellite will typically be in the 1–10 kg range (also this can be a cube-sat.) A pico satellite is in the 100 g to 1 kg range and a femto satellite is in the 10–100 g range (Definition of Miniature and Small Satellites; available online at: https://en.wikipedia.org/wiki/Miniaturized_satellite)

used for commercial business or machine-to-machine services. In such cases connection times are achieved in a matter of a few minutes.

In other cases, a small satellite constellation with a very large number of spacecraft (50–75 in number) might provide instantaneous mobile telecommunications as a completely commercial voice service for civilian and military usage. The latter is the type of small satellite constellation represented by the Iridium and Globalstar mobile satellite networks. Even these types of constellations might be of quite different design. In the case of the Iridium global satellite network for mobile communications, there were inter-satellite cross-links and coverage of the entire planet due to the use of polar-orbiting satellites. In contrast, the Globalstar constellation was designed to only provide service between latitudes 55° North and 55° South and was not designed to provide cross-links. This requires Globalstar to have more ground facilities to achieve interconnection.

Most small satellites are deployed in low Earth orbit, but it is possible to deploy constellations of small satellites in medium Earth orbit or occasionally even in geosynchronous or other special orbits. It is even possible to launch a small satellite into geosynchronous orbit for strategic purposes or unexpected purposes such as a military intervention or an emergency (e.g., epidemic outbreak).

Commercial and scientific experimental "small satellites" such as the NASA FAST satellite (which has a mass of 200 kg) can range greatly in size between 100 and 1,500 kg, and can also vary greatly in capability and functionality. Larger small satellites are usually designed with a full range of capabilities.[3] Thus, they are equipped with battery and solar array power systems, active stabilization, pointing and positioning, and a very important capability to either de-orbit or to be maneuvered to a so-called parking orbit. In the case of smaller satellites, capabilities may be much less. They may use a more crude gravity-gradient stabilization system. They may have no active or passive orientation system. They may operate with very low gain omni- or squinted beam antennas with no active stabilization or pointing capability at all. The smallest nano or cube-sat units have no active de-orbit capability at all.

1.2   Understanding the Differences Between Large and Small Satellites

The types of spacecraft primarily used for communications, remote sensing and Earth observation, meteorology, navigation, defense applications, scientific exploration and human habitation are typically large structures weighing thousands of kilograms. This is because there are economies of scale in most high technology devices, and satellites are no exception. For instance, a large solar array or battery can be more efficient than a smaller one when their respective capacities are measured in watts per kilogram. Also, the ratio of the mass of a payload of spacecraft

---

[3] Leonard David, Small Satellites Finding Bigger Roles as Acceptance Grows, *Space News*, Aug. 29, 2011; available online at: http://www.spacenews.com/article/small-satellites-finding-bigger-roles-acceptance-grows.

compared to that of the "bus" that carries the payload increases as a spacecraft becomes larger. Part of these economies of scale is achieved because the labor costs associated with designing and testing a spacecraft becomes proportionately less as size increases. A satellite that is four times larger might only cost a third more money to design, engineer, and test.

Yet, while perhaps 95 % of the total mass represented by all spacecraft in Earth orbit can be classified as medium- to large-scale spacecraft in terms of size, the number of satellites that are classified as cube, micro, nano or simply "small" satellites represents a much larger proportion in terms of numbers when compared to the total number of satellites in Earth orbit. A thousand cube satellites weigh just 1,000 kg (equivalent to just 0.25 % of the mass of the International Space Station, which weighs about 400 metric tons). Even the Hubble Space Telescope at 11,000 kg or perhaps a large-scale telecommunications satellite weighing 10 metric tons represents the equivalent of many thousands of microsatellites.

One might logically ask: If larger spacecraft are more efficient, why do we see so many of these smaller craft in space? Many different reasons account for this phenomenon. The motivations and objectives of different types of users vary greatly. In addition, new micro-miniaturization technologies are also making smaller satellites – particularly in lower Earth orbit constellations – more and more efficient. This can be particularly true when a constellation of small satellites is being deployed to accomplish a mission. The motive behind a small satellite can range from "nerdy fun" to a narrowly targeted experiment or verification of a new space technology to a state's national prestige in claiming to have built and launched a satellite to the deployment of a constellation of satellites to provide a commercial service, or even to the fulfillment of a governmental or military objective.

What we do know is that deploying too many of these tiny spacecraft can create an orbital debris problem, since it contributes heavily to the accumulation of too much junk – especially in low Earth orbit. Many small satellites heaved into space with a minimum of effort can stay there from 20 to 50 years or even more. So, as we address ways to make small satellites better and more capable in this book, we will also explore the regulation of these small spacecraft and new low-cost and simple technology that can assist their de-orbit once their original objective has been met. Others are advancing incentives and regulations that would consolidate small satellites experiments and projects. This would allow them to fly at low cost and conveniently on the International Space Station, on private space platforms, or perhaps as "hosted payloads" on larger spacecraft. The focus is on accommodating those who wish to design, build, and launch small satellites but do so in a way that minimizes the creation of space debris.

1.3   Various Types of Small Satellites and the Rationale(s) for Their Deployment

There are many space applications or research projects where the best solution for a particular mission is to launch a cluster of very small satellites as a constellation rather than one large satellite. Low Earth orbit satellites in a cluster can provide communications with only a very short transmission delay in contrast to a geosynchronous communications satellite that involves a ¼ s delay for an Earth-to-satellite or satellite-to-Earth channel and nearly a ½ s for a complete round trip circuit. Student educational and research projects must be simple and low cost in order to be launched within the constraints of university research budgets. Amateur radio operators need only a low power signal to complete a link. Developing countries or emerging economies want to be able to say that they have been able to design and build a satellite. International aid agencies need only to get simple text messages to rural and remote locations and thus do not need real time broadband communications. Defense agencies may need a specific ability to provide surveillance or communications links for a specific target area for only a short period of time.

The basic rationale behind small, micro, cube, or even smaller satellites is quite easy to understand. Such compact satellites are low in cost, easier to launch, and can open up new opportunities for students, small organizations, and experimenters who wish to have access to space. As a result of ever more powerful processors and application-specific integrated circuit (ASIC) devices, quite small satellites can carry out some rather sophisticated functions.

In short, there are many reasons for the launching of small satellites, but in many cases it is simply a matter of a limited budget for designing, building, and launching it. This can frequently add up to what is generically referred to as a "cube-sat." A conventional definition is a small satellite that is a $4 \times 4 \times 4$ in. (or $10 \times 10 \times 10$ cm) cube that weighs no more than 2.7 lb (1.2 kg). As one moves up from cube-sats, there is a range of options. There can be small satellites such as an Oscar satellite for amateur radio (typically in the range of tens of kilograms) up to scientific, experimental and even application satellites whose weight falls within the range of hundreds of kilograms. Everything is relative here. Today, a satellite weighing 1,000 kg can still be considered a small satellite in comparison to spacecraft of about 10 metric tons. Examples of small satellites include store and forward satellites or small remote sensing satellites such as might be typically designed by those organizations that specialize in small satellites; for instance, the Surrey Space Centre (particularly its commercial spin-off Surrey Space Technology, Ltd.,) or the Utah State University's small satellite program.

Another set of reasons is attributable to the concept of deploying low-Earth orbit or medium-Earth orbit constellations to support commercial services. This type of small satellite constellation is often designed to link up mobile users who seek to communicate via compact, low power transceivers. Such network designs benefit from the much shorter transmission distance – and thus reduced time delay and more modest "path loss" associated with low orbits. The trade-off, however, comes at a price. In order to achieve global coverage with a low-Earth orbit constellation, a lot more satellites are often needed. Three satellites in geosynchronous orbit can provide global coverage. Ten to fifteen satellites in medium-Earth orbit can also cover the globe, but there is a need to deploy something like 50–75 satellites to cover the world from low-Earth orbit.

The coverage concept here is as simple as recognizing that if you have "taller poles" or climb taller trees, this affords a wider view. In the case of a geosynchronous communications satellite network, one can deploy three large commercial satellites and achieve global coverage. This is because at that altitude – almost a tenth of the way to Moon – the satellites have a much higher and wider-scale view. Such a system, however, has the disadvantage of a much longer transmission delay, and a huge amount of path loss due to spreading of the signal from so far away (unless one has a gigantic antenna to concentrate the beam). There is, of course, a higher launch cost per satellite to send it into geosynchronous orbit, but the net over-all launch costs are reduced because there are much fewer satellites being launched.

The alternative, if you are seeking to create a global communications network, is to deploy a large number of smaller satellites in low-Earth orbit. These smaller low Earth orbit satellites have much less path loss (or spreading out of the signal as it goes from satellite to the ground or vice versa). The low Earth orbit constellation involves much less transmission delay and it generally enables the use of much smaller transceivers (or handsets) in the case of the widely distributed users.

1.4   The Rising Problem of Orbital Debris and Small Satellites

The problem of orbital debris is, in some ways, simple and, in other ways, quite complex. First of all, there are currently only eight spacefaring nations that have the capability to launch artificial satellites into orbit on a regular and consistent basis, and three others, namely Iran, North Korea, and South Korea that are perfecting their launch capability. Launches carried out by the United States, the U.S.S.R./Russia, China, and Europe are responsible for over 90 % of the total active and defunct satellites, rocket motors, and various elements in space – the so-called space junk presently in orbit.

Since October 1957, about 6,000 satellites have been placed in orbit of which about 1,000 are still operational. There are more than 21,000 objects tracked by the U. S. Space Surveillance Network 5–10 cm in low Earth orbit and 0.3–1 m in geo-synchronous orbit. Further, the total number of pieces of space debris in Earth orbit is increasing exponentially (See Fig. 1.2). According to a National Research Council Report, space debris has already reached a "tipping point,"[4] and an increasing num-ber of space satellites will further exacerbate the situation and thus threaten the sustainability of space activities in the long-term.

There is now legitimate concern that the so-called Kessler syndrome (where the amount of debris continues to cascade in a 'chain-reaction multiply out of control) might occur in the near future unless active debris removal processes are instituted and guidelines against new debris creation strengthened. The diagram in Fig. 1.3 below shows that the amount of space debris in Earth orbit has now increased to over 6,300 t, and the number of pieces unfortunately continues to grow.

---

[4] "Accumulation Is Past 'Tipping Point', NASA Urged to Clean it Up" Read more at http://planet-save.com/2011/09/06/report-space-junk-accumulation-is-past-tipping-point-nasa-urged-to-clean-it-up/#sBmzBtBGichedSy1.99.

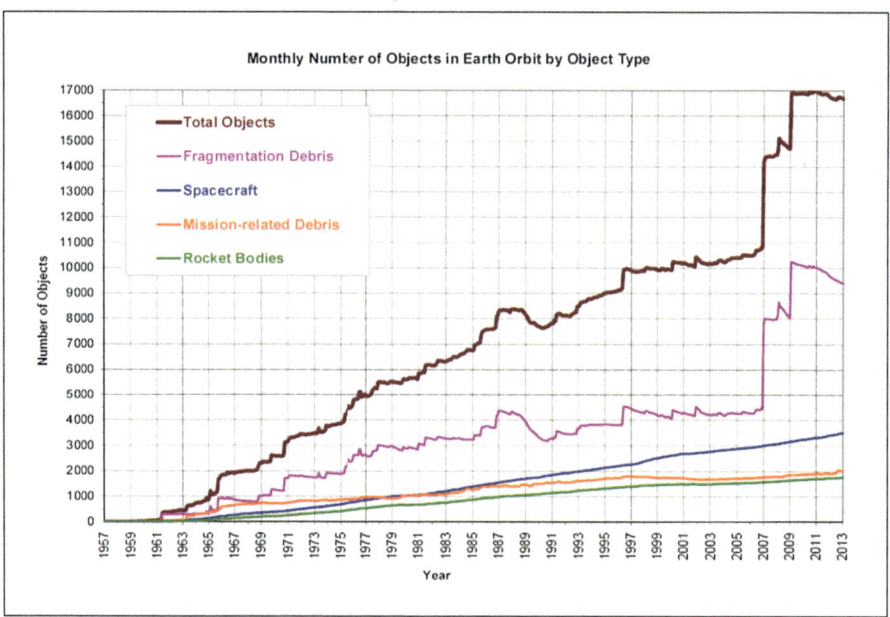

Monthly Number of Cataloged Objects in Earth Orbit by Object Type: This chart displays a summary of all objects in Earth orbit officially cataloged by the U.S. Space Surveillance Network. "Fragmentation debris" includes satellite breakup debris and anomalous event debris, while "mission-related debris" includes all objects dispensed, separated, or released as part of the planned mission.

http://orbitaldebris.jsc.nasa.gov/newsletter/pdfs/ODQNv17i1.pdf

Fig. 1.2 Historical evolution of the number of objects in space (Courtesy of NASA)

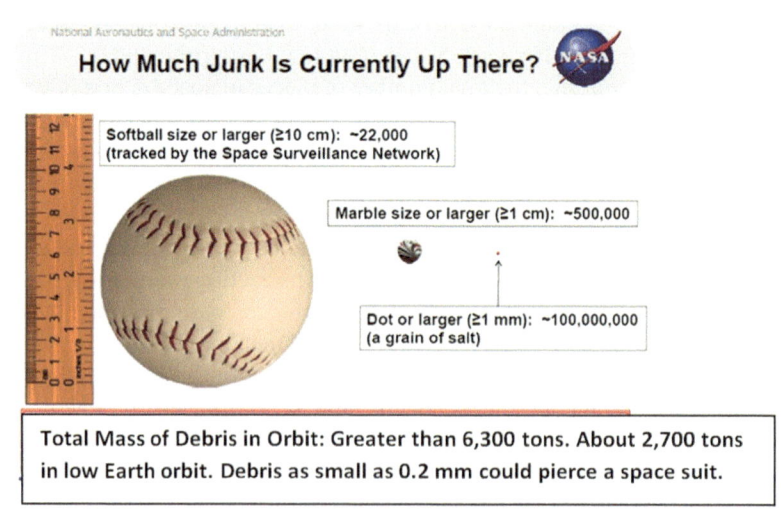

Fig. 1.3 Quantifying the overall space debris problem (Courtesy of NASA)

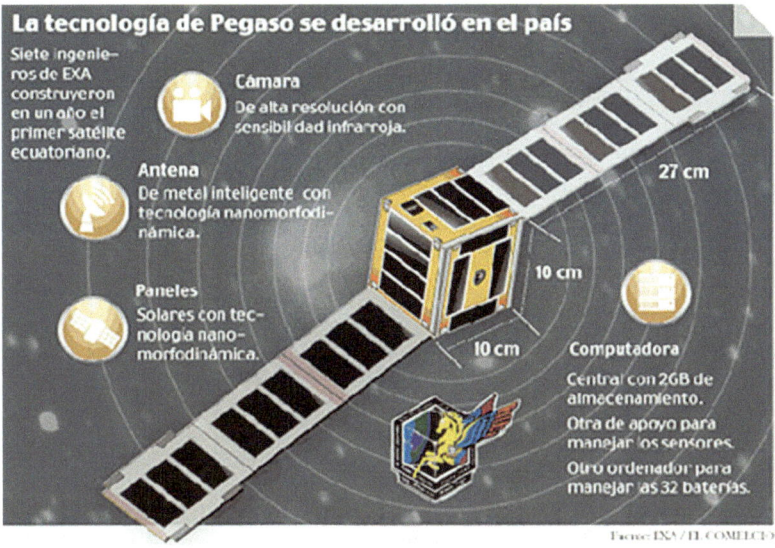

Fig. 1.4 Ecuador's satellite Pegaso (Pegasus) (Courtesy of Satnews, May 29, 2013) infra note 5

The first documented collision was in 1996 between a piece of launcher debris and the CERISE microsatellite – which survived and was returned to service. In recent years, there have been close encounters and some collisions with space debris. On February 10, 2009, a dead Russian Cosmos 2251 satellite collided with active Iridium 33 satellite, thereby destroying the latter, interrupting telecommunications services and creating thousands of pieces of space debris. Another dramatic collision occurred on May 23, 2013, between a piece of debris from a Russian rocket and Ecuador's first cube-sat called "Pegaso" (See Fig. 1.4).[5] This collision shows that even cube-sats are not immune from the dangers posed by space debris. According to NASA, even smaller objects are possibly hitting small satellites.[6]

Although the United States, the U.S.S.R./Russia, China, and Europe have launched satellites, space stations, and other space objects into space on behalf of other countries, these launching states as major spacefaring countries bear the greatest burden of responsibility under the current international space treaties to address this problem. The space debris problem is of vital concern, and the contribution of small satellites to this issue is thus one of the key areas that this book addresses. At this point, it is thought that the best remediation strategy is to find ways to actively remove large space objects, but finding a better strategy with regard to small satellites is still a matter of prime international concern.

---

[5] "Ecuador… Pegaso Spinning Like A Top… (Satellite)", *Satnews*, May 29, 2013; available online at http://www.satnews.com/story.php?number=529910663 Wang Ting, "With Ecuador's Only Satellite Hit by Russian Space Debris, Liability Should be Established," May 29, 2013; available online at: http://www.huffingtonpost.com/wang-ting/with-ecuadors-only-satell_b_3356479.html.

[6] NASA, Orbital Debris: Quarterly News, April 2013, p. 1.

1.5   The Concept and Scope of This Book

The concept of this book is to accomplish several key objectives within the constraints of a brief book that provides a higher level overview of the field and its regulation. First we seek to describe and define the various types of small satellites and explain who designs, builds, and/or operates them. We also seek to describe the key technologies that are of special interest and importance to small satellites and especially those new technologies that are currently evolving to help sustain the small satellite industry.

This book uniquely provides the historical background of the small satellite industry and, as well, discusses new solutions needed to support the future development of the sector. It explains the various reasons why it still makes sense to launch small satellites of various shapes and sizes today. It explains why a diversity of design and purpose meets the various needs of so many different types of users, and how new concepts such as "host payloads" are changing the small satellite industry. Finally this book takes on the issue that is now considered a major problem for the entire satellite industry but has special significance for small satellites – namely orbital space debris.

This book is intended to provide a broad and up-to-date overview of important aspects of the small satellite industry in simple, mostly non-technical language. It explores the technical, operational, financial and regulatory challenges now faced by this very special part of the global space enterprise. In the past few years, these challenges have entered a totally new phase for several reasons. One positive driver is the new economic incentives that derive from micro-electronics and new low-cost commercial launch capabilities that are now bursting onto the scene. Another driver – a negative one this time – is the problem of orbital debris. This problem has reached a new critical level of concern, but new technologies and hosted payloads may provide at least a partial answer.

This book further examines different types of applications for which small satellites of various types might be deployed and, as well, investigates the underlying reasons. It also explores innovations that are constantly being introduced. Aside from coming up with ways to achieve better performance with smaller satellites, these innovations also involve ways to minimize the problem of orbital debris that is increasingly getting worse. Examples of these innovations include: (i) low cost and effective ways to de-orbit small satellites after their missions have been completed; (ii) ways to consolidate modest space experiments and missions such as the "nano-racks" that allow a large number of experiments to be carried out on a specially designed rack that flies on the International Space Station; (iii) the innovative use of small satellite constellations or other satellites to serve as a "host" to other payloads that would otherwise be separately deployed as independently-flying small satellites; (iv) new deployment concepts and systems for small satellites; and (v) new low-cost commercial launch systems for small satellites.

There are now well over 20,000 major elements of orbital debris now being tracked, and low Earth orbit, where most cube, nano, micro, and small satellites reside is, by far, the most congested area. The challenge is thus to not only make smaller satellites more capable, cost effective, and versatile, but to seek other innovations that address the orbital congestion problem. Of course, not every innovation will necessarily contribute to solving the orbital debris problem. For instance, one recent innovation known as the JEM Small Satellite Orbital Deployer (J-SSOD) involves the addition of a new capability to the Japanese Experimental Module (JEM) on the International Space Station (ISS). The new capability consists of a mechanical space robot arm installed on the JEM for the purpose of "launching" small cube-sats from the International Space Station (ISS). This innovation will allow more of these cube-sats to be deployed at low cost after being transported to the ISS in a regular cargo supply mission.[7]

We have endeavored to present the latest research and evolving new technology that might aid and advance the de-orbiting of small satellites so as to minimize the ever-increasing orbital debris problem. Most small satellites tend not to have fuel or de-orbit mechanisms to allow or facilitate their return to Earth. As such, a key element of considered in this book relates to new types of solutions concerning de-orbit and end-of-life procedures. Finally we address the regulatory environment that governs the deployment, operation, and de-orbit of small satellites and the liability provisions related to this class of satellites.

1.6 The Structure of This Book

Chapter 1 has provided a quick overview of the field of "small satellites" and discussed, in a broad way, the various kinds and uses for such space objects that range in size from a few grams to 1,500 kg. Clearly, there are many types of these satellites used by a dynamic range of applications and scientific experiments. It also provides some insight into the problem of orbital debris and the increasing risk of an avalanche effect (known as the Kessler syndrome) that could create a deadly cloud of debris.

Chapter 2 provides much more detail with regard to various designers, builders, and operators of small satellites. Small satellite systems have been largely developed for the following reasons: (i) military and strategic; (ii) emergency and disaster relief and short wave amateur radio; (iii) educational and scientific applications; and (iv) start-up space programs for various countries. These key users and developers of small satellites will be discussed. Also, the two leading entities/programs that have taken a leading role in developing small satellites, namely the Surrey Space Centre and the Utah State University program, are likewise addressed.

Chapter 3 sets forth in the latest innovations in small satellite design, deployment and de-orbit capabilities. The chapter indicates the evolving technologies that have served to make small satellites more capable of being used for a wider range of commercial, scientific, and strategic applications. A special focus is given to new

---

[7] "JAXA Adds New Small Satellite Orbital Deployer to the Japanese Experimental Module (JEM) of the International Space Station (ISS), Via Satellite, January, 2013.

innovations such as hosted payloads, consolidated small satellite projects, and low cost systems to add the de-orbit ability of small satellites at end of life.

Chapter 4 presents an overview of the legal and regulatory background associated with space activities and how they relate to all forms of satellite deployment, especially small satellites.

Chapter 5 proceeds to explain in greater detail how different countries provide for the licensing, regulatory approvals, and frequency allocations associated with satellite deployment, including small satellites.

Chapter 6 explains the orbital debris and the consequent responsibility and liability that nations take on when they launch space objects under international law, and in some cases, under national law. This is a concern that grows in importance as this problem continues to grow. The increasing launch of small, micro, nano, and even femto satellites has not only given rise to concerns about the Kessler syndrome occurring much more rapidly than had once been thought but has also stimulated much more thought and action related to regulatory controls and processes. This chapter seeks to note what regulatory procedures are in place, and notes ongoing efforts to address the orbital debris problem and the liabilities that may be incurred.

Chapter 7 explores possible new solutions and innovations that can allow small satellites to be more cost effective, more efficient in achieving their mission objectives, or better able to avoid the problem of space debris and to mitigate any liability that may be incurred. Thus, this chapter seeks to explore what viable new solutions might be possible. Among other items, these proposed solutions include: (i) incentives to consolidate payloads; (ii) low cost end-of-life de-orbit systems; (iii) new regulations setting timetables for de-orbiting; (iv) regulations to use "green" station-keeping fuels or gravity-gradient booms; (v) revised space object liability provisions and a new definition of space debris; and, (vi) new arrangements for systematic global space traffic management.

Chapter 8, which is the concluding chapter, addresses the ten top things to know about or consider with regard to the design, launch, operation, and final disposition and de-orbit of small satellites.

Chapter 2
The Development of Small Satellite Systems and Technologies

2.1 The Evolution of Small Satellites

The fascinating world of small satellites began when Sputnik launched the Space Age in October 1957 and surprised the world with the knowledge that humans could lift artificial satellites into Earth orbit. But that was nearly 60 years ago. Soon the rockets that were developed in the U.S.S.R. and the United States were capable of launching payloads that were not tens of kilograms or tens of watts in size and power but represented thousands of kilograms and thousands of watts. The drive to launch larger and larger civil spacecraft was accelerated by several factors. These included higher power (solar cell arrays), large aperture and high-gain antenna systems, and, in time, the building of space habitats to house astronauts and large scientific instruments such as the Hubble Space Telescope.

As noted in Chap. 1, there are a number of economies of scale and scope that help the cost efficiencies of a large remote sensing or large telecommunications satellites to become much greater than small ones. These include proportionally lower costs of design, engineering, testing, and verification as well as launch. In the case of a telecommunications satellite, one large parabolic reflector can be used by a small multi-beam feed system to create dozens or even hundreds of spot beams to support intensive frequency re-use. Since the electronics are small, the driver of the mass and size of the satellite is the aperture of the communications satellite antenna.

Thus, for many years the predominant trend in commercial satellites was bigger, more capable, and more cost-efficient satellites. The larger launchers also tended to be more cost efficient in terms of cost of lift as measured in dollars per kilogram of payload. As a result of these efficiencies, the cost of a satellite telecommunications circuit since the start of service in 1965 has plummeted from $64,000 (in U. S. dollars) per month to about a dollar per month.

In light of the tremendous increase in lift capacity of today's launch vehicles, such as the increase from Atlas 1 to Atlas V or from Ariane 1 to Ariane 5, and the economies of scale and scope, one might jump to the conclusion that there is no longer a need or even a true market for small satellites. This is clearly not the case.

R.S. Jakhu and J.N. Pelton, *Small Satellites and Their Regulation*, SpringerBriefs in Space Development, DOI 10.1007/978-1-4614-9423-2_2, © Springer New York 2014

Today there are many more small, micro, cube, and nano satellites being launched than ever before. As noted in Chap. 1, the reasons why so many of these small satellites are being launched are numerous. We believe that one of the more useful ways to explain this ongoing interest in designing, building, launching, and operating small satellites is to break the market down by categories of users. Let's begin with military applications.

2.2 Small Military and Defense-Related Satellites

Small satellites for military applications can actually cover a number of needs. In light of the wide range of purposes now delivered by military space systems, only the citation of specific examples can provide a useful understanding of why small satellites might be appropriate or even best for certain strategic needs. One should start, however, by noting that the military has many ongoing requirements that depend on quite large satellites. Surveillance satellites and telecommunications satellites, for instance, can be the size of a small house. Thus, military satellites can be very large, large, medium, small, and micro-sized spacecraft, depending on the specific need. The various types include:

2.2.1 Rapid Deployment Small Satellites for a Specific and Newly Emergent Theater of Combat or Other Exigent Need

In the case of commercial or civil governmental services the demand for a particular service such as telecommunications, remote sensing, meteorological, navigational, geodetics, etc., is well established, and the transition from an existing satellite to the next generation with enhanced capability or capacity can be easily planned and executed. In the case of military systems, the outbreak of hostilities or an emergency situation prompted by a terrorist attack can occur with little or no warning. The military has adapted to such needs by having small and dedicated satellites that can be launched with little advanced warning. The military has also promoted the idea among satellite suppliers to have components of small satellites (i.e., antennas, power supply, processors, stabilization systems, and thrusters) that could be quickly assembled and launched on short-term notice. These innovations have helped others be able to order small satellites with much quicker delivery schedules.

2.2.2 Constellations for Mobile Communications or Machine-to-Machine Data Relay

The ability to communicate by voice, data, and even image to the "edge" of networks where combat soldiers or emergency relief operations are located is critical

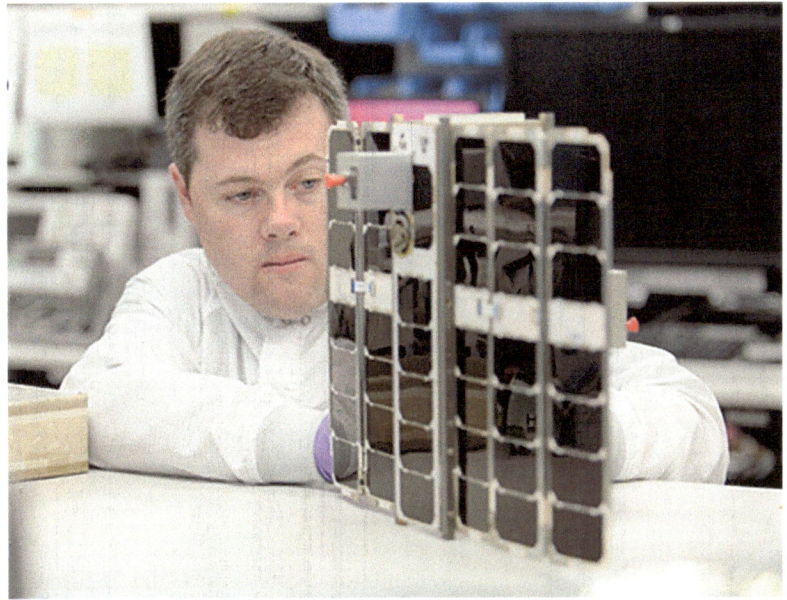

Fig. 2.1 One of the OR-3 nano sat units built by Boeing for the U. S. Air Force (Graphic provided by the Boeing Corporation)

to modern military activities. Several military operations are now relying extensively on the Iridium and Globalstar small satellite constellation for mobile communications services. The U. S. military is now deploying its so-called MUOS small satellite LEO constellation to support remote mobile communications on a global basis. There are negotiations underway with regard to the next generation of the Iridium small satellite network to place "hosted payloads" on this network to support specialized military services.

2.2.3   *Small Satellites for Collection of Data from Ground, Sea, or Other Distributed Sensors*

Many types of military services require broadband (i.e., television or videoconference type) service and large commercial or defense satellites meet these needs. There are other applications, however, that involve only the collection of very short data bursts from remote buoys or ground-based sensors to detect and monitor sea, snow, or other meteorological conditions. The U. S. Air Force under their so-called SENSE program is deploying and testing two 3U cube-sats (30×10×10 cm and weighing 4 kg) to evaluate whether the meteorological nano sats might be able to provide reliable weather data from around the world (Fig. 2.1).[1]

---

[1] Boeing [NYSE: BA] has delivered two *Space Environmental Nano-sat Experiment* (SENSE) satellites to the US Air Force Satnews Daily December 21, 2012; available online at: http://www.satnews.com/cgi-bin/story.cgi?number=789943063.

2.2.4   Experimental Packages to Test New Technology or Service Delivery Systems

Beyond the OR-3 program the U. S. Air Force has a Space Test Program (STP) that seeks to develop new technology to advance space capabilities in such areas as meteorology, surveillance, remote sensing or communications. The Advanced Research and Global Observation Satellite (ARGOS) was designed and launched by the Air Force Space Command in 1999. This is just one such test program. Although this system with multiple experiments might be considered too large to be a small satellite, earlier phases of the Space Test Program included smaller missions designed to test smaller experimental packages.

2.3   Commercial Constellations

The evolution of the satellite industry has unfolded in various phases. Initially, in the late 1950s, satellites were launched into low Earth orbit as the first limited rockets and launcher systems were barely capable of achieving very rudimentary orbital speeds, and stabilization systems were almost non-existent. By the early 1960s it was demonstrated that satellites could be successfully launched into geosynchronous orbit and operated there for extended periods of time. Geosynchronous satellites of greater and greater capacity with higher gain antennas and more capable sensors were deployed. By the 1980s, however, there was interest in deploying satellites for mobile communications, but special constraints emerged in the development of these services. These constraints involved the need to have low-gain transceivers for users on the ground, and also a desire to minimize transmission delay prompted thoughts of using a large constellation of small satellites to provide global coverage rather than having a few satellites in geosynchronous orbit. The Iridium, Globalstar, and Orbcomm systems were designed on this premise, and all envisioned dual-use applications to meet commercial, governmental, and defense communications requirements.

Today the U. S. defense network for mobile communications known as MUOS is also being deployed. This network to support mobile communications for the U. S. Navy, however, is a large-scale geosynchronous satellite. This design that features a very high gain deployable antenna represents a move away from using a low Earth orbit small satellite constellation to support defense-related communications. In the commercial mobile satellite industry, this same strategy has also been employed by Inmarsat with its high-gain geosynchronous satellite Inmarsat 4 and by the Thuraya system that has also deployed very high-gain geosynchronous satellites instead of using a small satellite constellation.

Similar low Earth orbit constellations may also be used to support meteorological, surveillance, and other applications, but their prime use is for mobile telecommunications. The satellites that constitute commercial constellations can be in different orbital constellations with different masses. For instance, Iridium satellites have a mass of 680 kg and are in 780-km high polar orbits within a constellation of 66 satellites plus spares. Globalstar satellites have a mass of 550 kg and are in a 1,400-km high orbit that are inclined up to 52° of latitude above and below the equator within a constellation of 40 satellites plus spares. The Orbcomm constellation that is used for machine-to-machine messaging includes a variety of satellites with masses that have ranged from 42 kg up to 115 kg. The original constellation had 36 satellites in it, and the current generation has 18 satellites.

2.4   Small Satellites for Educational and Scientific Applications

Perhaps the predominant application of small satellites is for educational projects and scientific experimentation. This can be for the most basic of nano satellite experiments that students at elementary or secondary schools might undertake under a structured competition sponsored in the United States by the National Center for Earth and Space Science Education (NCESSE), with these experiments flying up to the International Space Station as part of the NanoRacks LLC enterprise. On the other end of spectrum, there can be quite small but sophisticated experimental satellites designed by the world's leading universities or space agencies. In short, the range of sophistication, size, and mass in this type of satellite is enormous. A nanorack experiment designed by 12-year-old fifth-graders that seeks to measure radiation effects on yeast under the NCESSE program that flies on the International Space Station rather than as a free-flyer in space is one such extreme example.[2] Another example is the New Millennium Space Technology 5 project (See Fig. 2.2). This project consists of three micro-satellites (each 25 kg in mass) that have been measuring Earth's magnetic field since their launch on March 22, 2006.[3]

In the case of commercial applications where there are specific services being provided and an established market, it is clear that larger satellites that offer economies of scale with satellite manufacturing costs, launch costs, and operating expenses make very good sense. But in the case of student projects, scientific experiments and projects where only one small sensor in space is required, small craft make good sense – particularly if the small satellite can be launched as a very small and/or ancillary part of a larger launch mission. In the case of a Nanorack LLC experiment that flies as a part of an International Space Station resupply mission, the launch cost effectively becomes zero since NASA considers this as a part of its support to the Science Technology, Engineering and Math (STEM) initiative. However, some university-based small satellites, such as those designed by Utah State, Surrey Space Technology, Ltd., or the University of Texas, Austin, can be fairly sophisticated (See Fig. 2.3).

---

[2] National Center for Earth and Space Science Education, Student Spaceflight Experiments Program; available online at: http://ssep.ncesse.org/.
[3] NASA New Millennium Program, the Space Technology 5 project; available online at: http://www.nasa.gov/mission_pages/st-5/main/index.html.

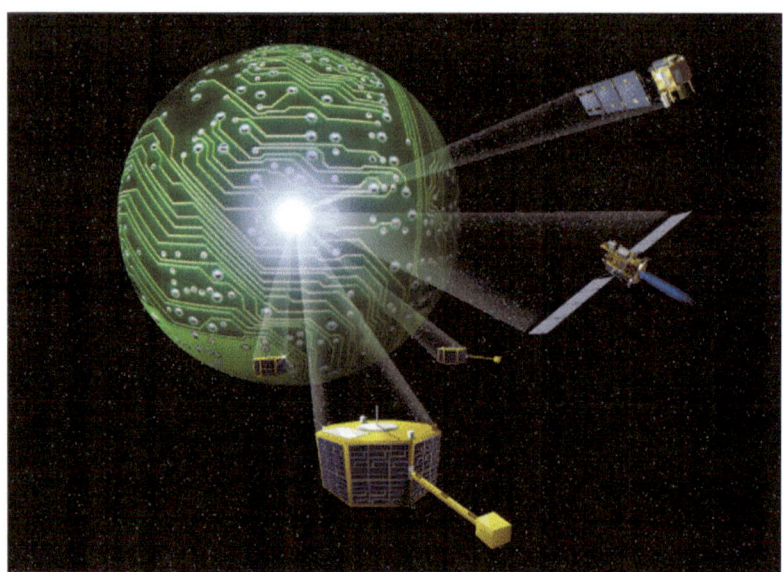

Fig. 2.2 NASA's small satellite new millennium program to measure Earth's magnetosphere (Courtesy of NASA)

Fig. 2.3 Frastac-A small satellite constructed at the University of Texas (Graphic Courtesy of the University of Texas, Austin)

Fig. 2.4 OSCAR 1, the first
of seventy such small
satellites launched into low
Earth orbit (Courtesy of
Amateur Satellite)

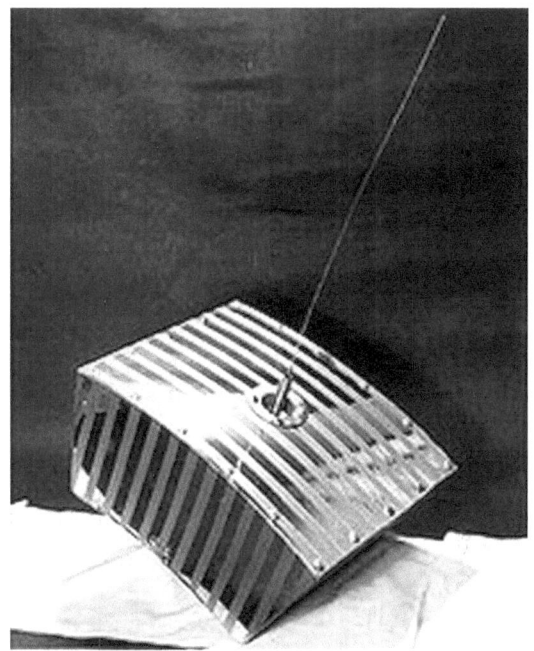

2.5   Small Satellites for Amateur Radio, Emergency, Disaster Relief and Other Social Applications

The other important application of small satellites can be to support emergency services, disaster relief, or medical or health services in very rural and remote areas where conventional communications or other services are not present. Livesat operated a two-satellite low Earth orbit messaging service to provide medical information as data relay on demand service. Small machine-to-machine data relay satellites that support everything from amateur radio to emergency or health services are within the resources of many organizations, particularly if volunteers can design and build the spacecraft and arrange for a low-cost launch.

Since the first amateur radio satellite known as OSCAR 1 (Orbiting Satellite Carrying Amateur Radio) was launched many decades ago, there have been over 70 of these satellites placed into low Earth orbit (See Fig. 2.4). These small satellites that have been designed and built by a number of countries around the world operate within the FM range of radio frequencies and is available to all "hams" worldwide.

2.6   Start-Up Programs in Various Countries with Fledgling Space Programs

Many countries that are just beginning a space program – or embarking on scientific measurement programs where a spacecraft is the optimal approach to take – typically embark on a small satellite program. In such cases, the satellite might be for various applications such as telecommunications, remote sensing, meteorological or navigational purposes. Alternatively, the small satellite may be for various types of space experiments. These small satellites may be built at universities or governmental research institutes. In a number of cases, there may be a partnership formed with one of the various groups that specialize in designing and building small state-of-the-art satellites. In Europe, the leading organization is Surrey Satellite Technology Ltd., which was associated spin-off company from the University of Surrey in the United Kingdom in 1985 as a commercial venture and is, in fact, now majority-owned by the European aerospace giant Astrium. The University of Surrey Space Centre continues with academic research into small satellite techniques.

When the Republic of Korea, for instance, started to design and build satellites they formed a cooperative relationship between KAIST (Korean Advanced Institute for Science and Technology) and Surrey Satellite Technology Ltd and the Surrey Space Centre to design remote sensing, telecommunications, and experimental satellites based on the very efficient Surrey small satellite platform. To date, scientists and engineers from Surrey Satellite Technology, Ltd., have been involved in over 40 small satellite missions involving Earth observation, imaging and space situational awareness, navigation, telecommunications, meteorology, military technical demonstration, technical verification and demonstration, and scientific experimentation. In 2003, SSTL formed the international Disaster Monitoring Constellation (DMC) of microsatellites. These projects have involved cooperative arrangements with Algeria, the Chilean Air Force, China, the European Space Agency, France, Korea, Malaysia, Nigeria, Portugal, Thailand, Turkey, Kazakhstan and even the U. S. Air Force.[4] In addition to other projects, SSTL are now building the 22 satellites in the European Galileo navigation constellation with its partner OHB in Germany. Another major small satellite center is at Utah State University. This university has the experience and ability to design and build small satellites. Other centers are evolving at the University of Texas, Austin, the University of Colorado, Boulder, and other universities around the world.

Countries that are starting up space programs and are not only designing and building spacecraft but also launching satellites have a wide range of options open to them. There is a growing range of alternatives, including dedicated small launch vehicles, ancillary payloads within a large-scale launch operation, or even by insertion into orbit from the International Space Station or other large space system. Finally, there is the option of becoming a "hosted payload" within another spacecraft program. In most cases, this would be in the case of riding onboard a constellation in low Earth orbit satellites.

---

[4] Surrey Space Centre and Surrey Space Technology Ltd. Historical Missions; available online at: www.sstl.co.uk/Missions/SSTL-Missions.

Chapter 3
The Technology of Small Satellites

The design and manufacture of small satellites can be broken down into two major categories of spacecraft bus and payload. A spacecraft bus is the platform that allows the spacecraft to support a particular function in space, and the payload is the hardware that is specifically designed to carry out the mission (such as telecommunications, navigation, Earth observation, meteorological sensing, surveillance or situational awareness, or some other form of space-related experiment or in-orbit testing of new technology). The bus must be able to provide the power; the thermal environment; pointing and stabilization; and the telemetry, tracking, and command (TT&C) capabilities needed to support the mission. The TT&C systems must have assigned frequencies to support the linking-up of the onboard systems with ground-based tracking and command signals as well as the relay of data to the ground to make sure the satellite is performing correctly.

The "bus" can be quite small, simple, and crude and thus supply very little functionality beyond power and perhaps some radio links to support command and data relay. It can also be relatively sophisticated even on a small satellite. There are buses even for small satellites that provide battery and solar power, heat pipes for thermal control, a tracking, telemetry, command and monitoring system, plus a system for stabilization and pointing of sensors, cameras, or antennas (which constitute the satellite's payload). There are organizations such as Surrey Space Technology, Ltd., Orbital Sciences Corporation, Sierra Nevada Corporation, as well as academic institutions such as Utah State University, the University of Colorado, Boulder, the University of Texas, Austin, etc., that are able to supply spacecraft buses to support a number of small satellite efforts. These start with simple nano-sats or cube-sats and go up to small satellites that can weigh hundreds of kilograms.

The payload of the small satellite, of course, defines its essence and mission. Small satellites typically tend to have a single instrument, sensor, or antenna system as its mission. This is particularly the case for a cube-sat whose typically dimensions are $10 \times 10 \times 10$ cm and which has a mass of about 1 kg (See Fig. 3.1).

The basic cube-sat, which is most commonly designed and built as a student learning experiment, is very simple in concept. There are solar cells on the outside

R.S. Jakhu and J.N. Pelton, *Small Satellites and Their Regulation*, SpringerBriefs in Space Development, DOI 10.1007/978-1-4614-9423-2_3, © Springer New York 2014

Fig. 3.1 A typical cube-sat
(Courtesy of NASA)

and typically lithium ion batteries to supply power, a simple antenna to support tracking and telemetry, and microprocessors and sensors or equipment to support a simple experiment. The basic miniscule cube-sat does not have the size or mass that is required to support any stabilization or pointing system and thus cannot be commanded. There are design efforts to employ on the larger 3U size cube-sat a solar array boom and a low power reaction wheel to provide some degree of stabilization and pointing capability.

Once a cube-sat is released into space, its final orbit is determined by its release. As such, a very low-gain "omni antenna" must be designed to send down telemetry and tracking data regardless of how the cube-sat is oriented in space. The payload might be a camera to snap pictures or a small Geiger counter or infrared sensor or other type of equipment to collect data about radiation, heat patterns, etc. A cube-sat is essentially a teaching device to allow aerospace students to learn some basic engineering concepts and skills and to realize the "thrill" of building a satellite that will fly in space.

The idea of a basic teaching exercise that cube-sats represents is broader than just a typical cube-sat configuration. There are kits that one can order online for "do-it-yourself" basic satellites that include electronics, computer processor(s), and dense data storage, and power systems. Such very basic satellites for student learning may or may not be in a classic cube-sat configuration and may be larger or smaller in size and/or mass. Such a typical kit can be expected to have the following types of components that can be configured into a do-it-yourself cube-sat. In addition, the kit normally includes a number of possible experiments or applications that could be accomplished with such a cube unit that might be on a half, full-scale, 1.5 scale, 2.0 scale, or even 3.0 scale size (See Table 3.1).

Table 3.1 Elements that might be found in a ready-to-build cube-sat kit

Ready-to-build kit for a cube-sat

- Complete, finished, and ready-for-launch cube-sat structure (in 0.5U, 1U, 1.5U, 2U or 3U size) with high strength, low mass, and large internal volume
- A Pluggable Socketed Processor Module for in-lab development and testing, a mother board and for the actual flight model a Pluggable Processor Module (PPM)
- Low-power, high-performance electronics based on your choice of PPM, using
  - 16-bit or 32-bit ultra-low power microcontroller
  - 8-bit or larger mixed-signal MCU
  - 16-bit high-performance microcontroller
  - 16-bit digital signal controller
- Multi-tasking software for the processor and a relevant software library
- Plug-in modem/transceiver support and built-in USB 2.0
- USB debug/Flash emulation tool (FET) for programming and debugging
- Power supplies (solar cells & lithium batteries), programming adapters, cables and tools

3.1 Technology Associated with More Sophisticated and Mission-Driven Small Satellites

An insightful publication entitled "The Future of Small Satellites"[1] provides a basis for assessing the ability of small satellites to achieve characteristics that are desirable for safe operation. An important contribution to that volume estimates attainable capabilities based on size.[2] There is a lot more to the small satellite world than just cube-sats and nano sats for students to learn about spacecraft design, to carry out simple experiments, or to test new materials or biological agents in a low-gravity environment. There are many larger and more sophisticated small satellites that can be designed for real space missions. Here, the technology associated with more sophisticated small satellites continues to evolve quickly. The relevant technologies can be usefully examined and discussed under the following categories: power systems; thermal control; ground surveillance and communication characteristics; stabilization and pointing systems; tracking, telemetry and control, maneuverability, etc.

3.1.1 Power Systems for Small Satellites

Power systems for small satellites are, in many ways, parallel to those employed in larger satellites. There are many options in terms of power systems for small satellites. These involve trade-offs between lower cost and lower performance systems

---

[1] Small Satellites: Past, Present, and Future, Henry Helvajian and Siegfried W. Janson, Eds., ISBN 978-1-884989-22-3, 2009.

[2] Siegfried W. Janson, Satellite Scaling Issues, p. 771, in Small Satellites: Past Present, and Future, Henry Helvajian and Siegfried W. Janson, Eds., Aerospace Corporation Press, 2009.

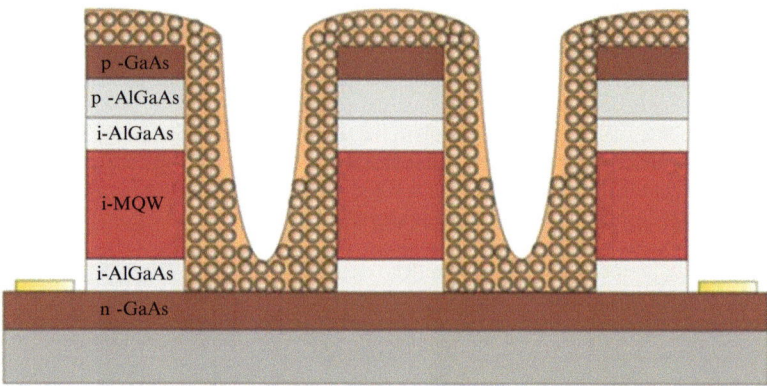

Fig. 3.2 A close-up of the "bumps" in a multi-junction quantum dot solar cell that would produce higher levels of electrical energy and at very high efficiency levels

versus higher cost and higher performance systems. These options include amorphous silicon and structured silicon solar cells and range up to higher cost multi-junction gallium arsenide cells capable of capturing energy in the high-energy ultraviolet range. In the future, there is the prospect of quantum dot technology. These quantum dot solar cells might be able to achieve perhaps 70 % efficiency in converting solar energy into electrical power for spacecraft use. This technology involves creating more effective surface exposure and more photovoltaic junctions to capture more solar energy across the spectrum. Thus the quantum dot solar cells would derive power from the most energetic ultraviolet range of solar radiation down through the visible spectrum. This technology is perhaps some 5–8 years away from commercial manufacture at viable cost levels (See Fig. 3.2).

There are also more efficient solar array and lower mass systems that involve thin film array systems that can be rolled out as opposed to deployed as rigid structures. Of course in the most compact and miniaturized small satellites, the solar cells are confined to the body of the satellite, and no solar arrays are deployed. Such a small satellite is limited in its power generation in that only about 40 % of the body would be able to receive solar radiation since the rest of the spacecraft would in effect be in eclipse. Solar arrays that can be deployed from a three-axis stabilized spacecraft have the advantage of tracking the Sun for maximum illumination. But, of course, such stabilization systems and the need for fuel to power the stabilization and pointing thrusters add weight to the satellite.

Another technology that can be utilized is a solar concentrator that serves to concentrate solar energy so that the solar array "sees" the equivalent of more than one Sun. Relevant research in this area is still seeking reflective materials that are lightweight enough to make such solar concentrators cost efficient. Currently, most small satellites use lower cost silicon solar cells and do not use solar concentrators. There is no systematic approach in this aspect of small satellites. Commercial missions such as mobile satellite constellations will typically use sophisticated solar arrays with high performance gallium arsenide solar cells. The same can be true for

Fig. 3.3 Nano satellite power and energy storage module (Courtesy of ISIS Cube-sat Solutions)

sophisticated small satellite systems designed by a governmental space agency. In contrast small experimental or student satellites will likely use much lower cost amorphous silicon solar cells.

Area per unit volume is greatest for spheres and increases inversely with object size. Therefore, solar-energized small satellites can have higher power to mass ratios than large satellites. However, the power attainable is still rather small. The potential is for no more than 10 W for body mounted cells on a nano sat deployed in a typical low Earth orbit – allowing for eclipse periods. This power output might be doubled if extensible panels are used. However, extensible panels add to mass and increase complexity and failure modes. Current standards and political constraints preclude nuclear energy sources in Earth orbit. This is particularly the case for low Earth orbit since budgets would normally exclude use of radioactive isotopes on small satellites in any event.

There is also the issue of energy storage for the time when the small satellite is in eclipse and no solar illumination is available. Considering allowable charge and discharge rates, nano-sats could sustain 1 W of continuous power for only a few months and as much as 10 W for a few days. In some instances of small satellite design where a particular experiment or test of a new space system or material does not require continuous operation, a lower weight and more compact battery can be employed. Such a battery storage system would thus be designed to provide only sufficient energy storage in order to support TT& C data relays rather than the operation of the payload during the eclipse period. Today, the cost of lithium ion batteries that have relatively dense storage capability has declined on account of their use in support of truly high volume market applications such as laptops, cell phones, etc. The research and development of technology by the most advanced research laboratories for the largest and most sophisticated spacecraft can often be efficiently transferred to smaller-scale projects (See Fig. 3.3 above).

Small satellite programs closely monitor research carried out in support of the most sophisticated programs to see if the outcomes can be usefully applied in smaller projects. If one examines the basic architecture of a large and massive satellite, it becomes readily apparent that the most significant elements of the satellite that are responsible for its large size/mass are usually its power and antenna systems. The first satellites launched into orbit had a power generating capacity of only a few watts. Today, there are massive communications satellites that might be generating 12–18 kW of power, and the solar array systems of the International Space Station can generate hundreds of kilowatts. High-gain antennas that are on the largest contemporary commercial satellites can be up to 22 m in diameter and weigh many hundreds of kilograms. These represent the other major driver of satellite size and mass. Indeed power and power systems are truly the principal drivers that make telecommunications spacecraft larger. Advances in electronics and optical processors, in contrast, keep shrinking the size and mass of modern spacecraft.

3.1.2   Thermal Control

A small satellite has a need for reasonable levels of thermal control so as to not overheat or overcool the electronic systems and the sensors or devices associated with the payload. Since small satellites are reasonably compact, the approach to thermal control is often based on the use of passive systems such as gold foil to reflect solar radiation to avoid overheating and enough absorptive materials to prevent the satellite from becoming too cold. Figure 2.2 above depicts the Fastrac small satellite, and this photo shows the reflective gold foil that serves to create the desired balance of solar heat reflectivity and heat absorption. It is possible that the design of reflective materials on the outside of a small satellite does not provide sufficient thermal conditioning necessary to support sensitive electronics inside the spacecraft. In the case of small satellites ranging up to 1,000 kg in mass, heat pipes to dissipate heat from the interior of the satellite may be required.

One of the effective solutions is what is called a miniaturized loop heat pipe (mini-LHP). Such a mechanism can provide an effective heat transfer function without many of the restrictions of conventional thermal control measures. Traditional techniques such as thermal straps and shunts, conventional heat pipes, mechanically pumped loops, and so on are not usually designed for small satellite use. If such techniques are used in small satellites, they could impose large mass penalties and exceed the weight budget for the mission. Such large-scale systems could also complicate system integration and create difficulties or complications with pre-launch tests – especially at the systems level. Swales Aerospace is one company that has developed a miniature multiple evaporator multiple condenser loop heat pipe that is scalable and is thus particularly optimized for use in a small satellite.[3]

NASA's New Millennium Program Space Technology 8 has developed a miniature loop heat pipe (MLHP). The complete miniaturized system has a mass of just over 300 g, or about a third of a pound. The European Space Agency (ESA), the Japanese Space Agency (JAXA), and other space programs have also devoted resources to developing miniaturized loop heat pipe systems with miniaturized

---

[3] Ahmed Habtour and Michael Nikitkin, "Miniature Multiple Evaporator Multiple Condenser Loop Heat Pipe"; available online at: http://digitalcommons.usu.edu/smallsat/2005/all2005/131/.

condensers as well.[4] Since the functions performed by such thermal control systems can be critical to the mission in terms of the operation of payload and spacecraft bus electronics, the objective of miniaturization must not overlook the need to achieve a high degree of reliability.

3.1.3 Ground Surveillance and Communication Characteristics

The laws of physics indicate that the aperture size used for imaging or remote sensing clearly limits the amount of electromagnetic energy that can be captured by a satellite. The image resolution obtainable by a small satellite depends on aperture size, and clearly in the case of a small satellite the antenna size cannot be very large. Larger aperture resolution can be simulated with the use of multiple, phase-matched small apertures on multiple small antennas flying in a close and fixed pattern. There are still penalties that occur in such a case. There are losses in terms of spatial frequency content of the scene and the amount of energy that can be captured (i.e., the signal-to-noise gain that is achievable for each aperture). In short there are severe limits on the amount of remote-sensing data and resolution that can be attained by a single nano sat or even a close flying constellation.

Communication antennas have comparable constraints. The tradeoff between antenna gain and effective isotropic radiated power is important. Using the nominal 10-W continuous power level estimate, a nano-sat in low Earth orbit could support a transmission rate of hardly more than one megabit per second or a few kilobits per second from geosynchronous orbit.

3.1.4 Stabilization and Pointing Systems

These two aspects of small satellite operation are not independent, and they impose different technical demands. Large satellites have high inertia, requiring larger torques to initiate motion and to sustain acceleration. Applying torques to the least massive elements of the system thread involved in re-directing bore-sights dynamically can mitigate this. Pointing components can take advantage of stable platforms whose stability is assured by the mass and inertia of the platform. Small satellites do not enjoy that advantage. Pointing and stabilization are very closely coupled. Stabilization is the most important element, since the satellite cannot be allowed to tumble. The low inertia allows high angular acceleration, which must be dampened.[5]

---

[4] J. Ku, L. Ottenstein, D. Douglas, "Multi-Evaporator Minature Loop Heat Pipe", NASA Goddard http://www.ntrs.nasa.gov/archive/nasa/casi.ntrs.nasa.gov/20080032843_2008031434.pdf.

[5] Samir Ahmed Rawashdeh, Passive Attitude Stabilization For Small Satellites, unpublished thesis submitted in partial fulfillment of the requirements for the degree of Master of Science in Electrical Engineering in the College of Engineering at the University of Kentucky, 2009.

Achieving sufficient control over a small satellite is always a challenge. Active techniques, which expend energy either in terms of propulsion or electromagneti- cally, employ actuators such as momentum storage devices. Such techniques can achieve bore-sight stabilization on the order of very accurate milli-radians.[6] However, active techniques may be a bit too excessive for mission-oriented nano satellites. Passive methods include passive magnetic stabilization, aero-stabilization, and grav- ity gradient stabilization. Passive techniques can achieve stabilization but with com- paratively less precision. Large satellites can, of course, do much better because of their ability to carry much more sophisticated pointing and stabilization systems.

One of the most important differences between a cube-sat or nano satellite and more capable small satellite in terms of mission capability and design is with regard to stabilization and pointing systems. A classic cube-sat, once released, cannot be controlled and remains in its release orbit until gravitational effects cause it to burn up in the atmosphere on its descent. Recently, there have been developmental efforts to create, for a 3U version of a cube-sat, the added capability of a solar array that could act as a gravity gradient boom and also to design a very low power reaction wheel that could achieve some degree of stabilization and pointing capability.[7]

Certainly a small satellite above the class of a cube-sat would typically have some means to orient itself in orbit and would thus be able to exert some degree of control as to its pointing. This capability may extend beyond a gravity gradient boom (or booms) and perhaps will have stored fuel and a thruster system to assist not only with its operation but also with active de-orbit maneuvers at the end of its life.

Perhaps the simplest means of stabilization is known as a gravity-gradient boom system that employs Earth's gravitational effect on deployed booms to generally "point" the satellite toward the ground below. This approach was employed fairly early in the development of satellite technology. The NASA Applications Test Satellite (ATS) series, and in particular, ATS 2, 4, and 5 used this stabilization technique. The medium Earth orbit ATS 2 was launched on April 6, 1967, and remained in orbit for 2 years. ATS 4 and 5 also employed this same technique of extending booms from these spinner spacecraft to achieve stabilization. There are a number of sophisticated small satellite missions that can and do use gravity gradient stabilization where exact point- ing is not required. Over a period of 2 years, the Orbview 1 (once known as Matlab 1) and pictured below (See Fig. 3.4) carried out in orbit testing for NASA's lightning detector sensor as a prelude to designing such sensors for the latest NOAA satellites.

The use of gravity gradient stabilization makes a good deal of sense for small satellites since much less hardware and no fuel is required, and they are relatively easier to construct and test. For these reasons, gravity gradient stabilization is less costly – although it also less accurate than active attitude-control systems. Gravity

---

[6] Siegfried W. Janson, Satellite Scaling Issues, p. 796, in Small Satellites: Past Present, and Future, H Helvajia and S.W. Janson, Eds, Aerospace Corporation Press, 2009.

[7] Erich Bender, "An Analysis of Stabilizing 3U Cube-sats Using Gravity Gradient Techniques and a Low Power Reaction Wheel"; available online at: http://digitalcommons.calpoly.edu/cgi/viewcontent.cgi?article=1035&context=aerosp&sei-redir=1&referer=http%3A%2F%2F.

Fig. 3.4 Orbview 1,
small satellite with gravity
gradient boom extended
(Courtesy of NASA)

gradient stabilization as shown in Fig. 3.4 is the only passive attitude-control method
used for satellites.

This method of stabilization relies on the highly asymmetrical satellite mass distribution. There is a change in gravitational attraction as the orbit of a satellite
increases. The gravitational attraction in geosynchronous orbit, for instance, is 50
times less than it is at Earth's surface. When a satellite is equipped with a long
boom, this results in a change in its gravitational attraction, since the principal axes
of the satellite are no longer aligned with the orbital reference frame, and this creates a torque. Due to the asymmetric nature of the satellite, the spacecraft will experience a torque tending to align its axis of least inertia with the field direction of
Earth's gravity. However, the relative values of the satellite's moment of inertia
around the overall center will not only point the spacecraft toward Earth but will set
up a slight oscillation. Dampers must therefore be installed on the satellite to reduce
this oscillation. As one moves from the smallest to larger satellites in the hundreds
of kilograms size range, most missions will transition to the use one of the active
stabilization control methods indicated in Fig. 3.5 below.

The actual approach used for active stabilization control hinges on many factors,
such as the pointing accuracy required for particular missions, the overall mass budget, the desired mean time to failure for the satellite in terms of its expected lifetime
in orbit, as well as other factors. Currently, spin stabilization is not often used
because three-axis body stabilization affords greater pointing accuracy and allows
solar arrays to be constantly oriented toward the Sun to give 100 % illumination
versus the 40 % illumination typically associated with spin-stabilized spacecraft.

Reaction wheels are probably the most common choice for the larger class of
small satellites. This is because of proven reliability, reliance on electric power

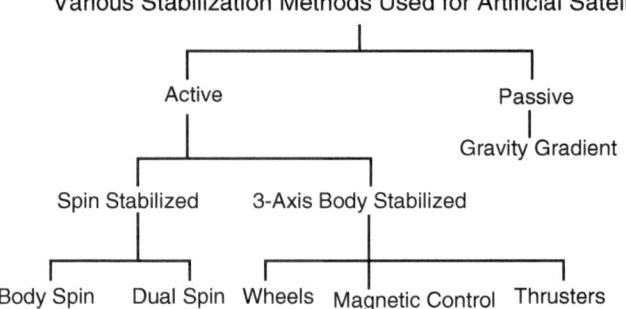

Fig. 3.5 Breakdown of various stabilization techniques for spacecraft

rather than fuel, and scalability of reaction wheels to spacecraft size. Inertia, momentum, or reaction wheels use the same principle of the kid's toy known as a top. The spinning of a wheel or more than one wheel in different planes can serve to keep a satellite oriented in a single direction. Reaction wheels for large spacecraft can spin at very high speeds of up to 5,000 revolutions per minute, but smaller reaction wheels for smaller spacecraft can spin at lower speeds and require much less electrical power to maintain these velocities.

Another issue is how does the satellite know where to point in space if there is no clear up or down? Here again, a number of options are available. One option involves the use of simple Earth, Sun, or star sensors that assist the satellite to point itself correctly. There are also now radio frequency beacons that allow more accurate pointing of satellites with a precision as accurate as 0.05°. In the case of satellites used for astronomy or for telecommunications, where spot beams must be aimed with great precision, such a high level of pointing accuracy is very important. For such missions, three-axis body-stabilized spacecraft are really the only viable option currently available.

3.1.5 Tracking, Telemetry, Command, and Monitoring

Yet another critical element of small satellite design is its tracking, telemetry, and command (TT&C) system. At least two things must be accomplished in order to operate a small satellite and derive useful data from it: (1) It must be possible to obtain accurate ranging data from the satellite in order to know where it is in orbit and to track its orbit with reasonable precision. (2) There must also be a transmission path in a suitable radio frequency band in order to obtain data from the satellite payload in a suitable downlink and, as well, to send commands and signals to the satellite so that it can start experiments, reposition itself, switch on backup units, or otherwise carry out essential functions. These tracking, telemetry, and command functions are carried out in radio frequency bands that are separate and distinct from those used by telecommunications, navigational, or radar remote-sensing satellites. Such missions will have specific radio frequency spectra assigned for their individual functions.

The antennas used to support TT&C functions are small, typically conic-shaped low-gain systems. Low-gain antennas can be used because the data rates involved are not necessarily very high. But, even more importantly, the key is to have antennas that are capable of receiving a signal from virtually any angle in case something should go wrong and the satellite should fall into a tumbling motion or a flat spin. An "omni" antenna may have low gain, but it will pick up a signal from virtually any angle.

3.2 New Technologies to Protect the Payloads on Small Satellites

The challenge of small satellites is to launch a meaningful payload to carry out a useful mission within the small power, mass and size budget that such a platform provides. Fortunately, electronics and processors have continued to shrink in mass and size over time as large-scale integration and application-specific integrated circuits (ASIC) have allowed scientists and engineers to do more with less. Constellations of satellites in low Earth orbit working as an integrated network have also allowed many useful applications for small satellites to evolve as well.

3.2.1 Higher Gain Antennas

One of the bigger challenges has been to incorporate higher gain antennas on small satellites, particularly with the advent of phased-array antennas. A phased-array antenna, phased-array antenna system, or phased-array antenna feed system can be employed in the design of the payload for a small satellite constellation. This technology can be used to electronically form spot beams that create a more efficient telecommunications satellite. One example of this approach is the Iridium satellite system (See Fig. 3.6). The payload design deployed three phased-array antenna panels that allowed a relatively small Iridium satellite to create 48 spot beam patterns on Earth below. The 106 radiating elements allowed 16 beams to be created from each of the three antenna panels for the total of 48 beams. Since the panels were flat and did not have to be deployed in a parabolic shape (as the beams were electronically simulated), the satellite could be much more compact.

On a much smaller scale, Surrey Satellite Technology, Ltd., and the Surrey Space Centre collaborated to launch the STRaND 1 smartphone satellite into space (See Fig. 3.7). This satellite was deployed within a 3U cube-sat platform. The entire "satellite" weighs only 3.5 kg and has been tracked by amateur radio operators from around the world. Miniaturization was used throughout both the bus and the payload.[8]

---

[8] "Smartphone satellite "STRaND-1 Operational in Orbit" SSTL News, March 7, 2013; available online at: http://www.sstl.co.uk/News-and-Events?story=2132.

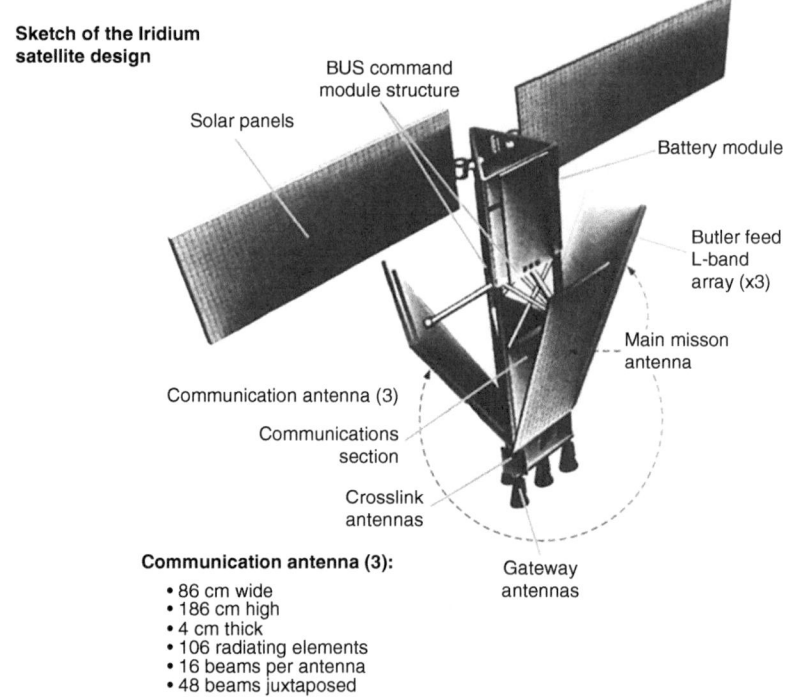

Sketch of the Iridium satellite design

BUS command module structure

Solar panels

Battery module

Butler feed L-band array (x3)

Main misson antenna

Communication antenna (3)

Communications section

Crosslink antennas

Communication antenna (3):
- 86 cm wide
- 186 cm high
- 4 cm thick
- 106 radiating elements
- 16 beams per antenna
- 48 beams juxtaposed

Gateway antennas

Fig. 3.6 Phased-array antennas on the Iridium satellites

In this case, cheap smartphone electronics is used to control the satellite. STRaND-1, which was built in only six months and as a training project between SSTL and SSC carries an amateur radio AC.25 packet radio downlink that operates at 437.568 MHz. It is able to transmit from its micro antenna at a bit rate of 9.6 kilobits per second using frequency shift keyed modulation and special NRZ1 encoding to maximize throughput. Information on how to receive and decode the downlink telemetry is available on the AMSAT-UK website. Here, the key to accommodating the payload's mission on a 3U cube-sat was micro-electronics and encoding technology rather than innovative antenna design.

Many payloads on small satellites are likely to involve sensors of some type. In this area microelectronics, applications-specific integrated circuits, miniaturized cameras and light and energy sensors, spectrographs, etc., can allow a compact payload to be accommodated on smaller satellite buses than was possible a decade ago. Further, many of these payloads require less power than they did a decade ago. This is generally the case of passive sensing systems for remote sensing and meteorological or Earth observation, but there is a major exception. Active sensing systems, namely radar satellites that must generate power to beam down, still require a major power source. As such, these types of "active sensing" devices require both major power supplies and thus large spacecraft. As noted earlier, improved multi-junction

Fig. 3.7 STRaND 1
Smartphone satellite
designed by Surrey Satellite
Technology, Ltd

solar cells, quantum dot technology, low-mass solar concentrators as well as improved and more dense battery storage such as lithium ion systems have certainly served to reduce the mass to power generating and storage ratio. Yet, these improvements can only go so far to reduce mass and size requirements. And there are always tradeoffs. A power system might be designed to generate, say, 25 % more power and store it for the same amount of mass and volume, but the cost of doing so in terms of more expensive technology might not result in the realization of significant overall gain even where reduced launch costs are taken into consideration.

3.2.2 Technical Advances to Consolidate "Small Satellite" Missions and Experiments

In general, large launchers are more efficient than smaller rockets. Similarly, larger satellites are more efficient than smaller ones. If there is one telemetry system to support twelve missions, rather than 12 telemetry systems to support 12 different small satellites, the greater efficiency of the former is clear. Consolidation of elements that constitute a satellite "bus" (whether it is solar arrays, batteries, thrusters for stabilization, thermal control systems, sensors for pointing and orientation, or tracking, telemetry, and control) almost always leads to efficiency gains. This also translates into lower labor costs for maintenance of a satellite in orbit. Thus, significant efficiency gains can be achieved by consolidating a number of small payloads that are designed for a particular mission in space. In spite of the consolidation, small independent space missions may still retain their own unique identity. One such approach is provided via NanoRacks, a company that lists the following "space firsts" on its website[9]:

- First company to own and market its own hardware on the space station
- First company to coordinate deployment of a satellite from the ISS
- First company to own and operate the External ISS Platform
- First self-paying high school space project
- First electroplating in space
- First terpenes in microgravity research
- First national space STEM program with no NASA funding (the National Center for Earth and Space Science Education that also works with the Arthur C. Clarke Foundation)
- First Vietnamese satellite in low Earth orbit (FPT University of Hanoi)
- First Israeli program on station (Fisher program)
- First Saudi program on station (KACST)
- First commercial payload on SpaceX (Multiple)
- First company to place customers on all ISS-related launch vehicles – the space shuttle, Soyuz, Progress, ATV, HTV, and SpaceX

Currently the NanoRacks Corporation advertises on its website the following services: (i) internal payloads that allow a series of experiments to fly to the ISS as "nano missions"; (ii) deployment of satellites from the ISS that range from cube satellites to larger small satellites; (iii) access to an external platform on the ISS for experiments and tests in a hostile space environment or for deep space observation; (iv) deployment opportunities from suborbital to deep space.[10]

By acting as a consolidator, NanoRacks allows a large number of tests and experiments in space to occur on a consolidated basis. Although NanoRacks is also involved in the deployment of separate cube and small satellites from the ISS and via other means, the main purpose is to be a consolidator and to minimize the number of separate missions that fly.

NanoRacks is not alone in this effort. Bigelow Aerospace is offering private companies and government agencies the commercial opportunities to fly experimental missions on its private space habitats for periods ranging from a few weeks to many months. There are also plans by JP Aerospace to create a lighter-than-air

---

[9] The NanoRacks Corporation; available online at: http://nanoracks.com/.

[10] Nano Racks Corporation capabilities; available online at: http://nanoracks.com/products/beyond-iss/.

Dark Sky Station that could fly experiments tens of kilometers above Earth. Other organizations, such as IOS systems, have indicated plans to fly people and experiments up on a commercial basis, and most of those commercial ventures that are planning to offer suborbital flights to passengers could also accommodate experimenters as well. One advantage of all of these various efforts is that the experiments would go up and then come back down without creating new space debris.

There are other options to provide consolidation of space missions and to reduce space debris that take an entirely different tack. One such approach that has become quite popular because it can reduce design, testing, manufacturing, deployment, and operating costs is the concept of "shared" or "hosted payloads." In 2011, a Hosted Payloads Alliance (HPA) was formed to create a mechanism for more effective communications between private enterprise and governments on possible sharing of missions and to explain more broadly the advantages of sharing payloads.

Today, large space service companies such as Intelsat, Inmarsat, SES, and Iridium have staff, and in some cases, entire offices dedicated to developing commercial arrangements with regard to hosted payloads.[11] Initially, the concept involved just one type of experiment, such as CISCO's experimental Internet Router in Space (IRIS) payload on an Intelsat satellite.

More recently, projects are being developed that involve a large number of payloads that can fly on a constellation such as on the next generation of Iridium mobile satellites (i.e., Iridium Next). In fact, one such major hosted payload project is now under contract. Iridium LLC has formed a joint venture with NAV Canada[12] to equip its next generation of mobile satellites with 50-kg packages (drawing some 50 W and up to 200 W of peak power) for an aircraft tracking capability.

Known as Aireon, this joint venture forms part of the replacement constellation for the Iridium global mobile satellite network. The Aireon system will "ride" on this new 66-satellite global airline tracking system. The stated goal is for the Aireon service to use space-qualified Automatic Dependent Surveillance-Broadcast (ADS-B) receivers to provide an unprecedented ability to track aircraft on a totally global basis. The receivers will normally operate at 100 kilobits per second but will be capable of supporting 1 megabit per second speeds if required. This joint venture will, for the first time ever, provide air navigation service providers (ANSPs) the capability to continuously to track aircraft anywhere in the world in near-real time, including over oceanic, polar, and remote regions.[13]

All of these innovative efforts that involve more efficient packaging and seek to put "small satellite" missions onto operations that can fly up and then fly down without creating orbital debris are welcome efforts. The fact that, in most cases, these consolidated space programs lead to cost savings in terms of design, testing, manufacture, launch, and ongoing operations helps to create the right incentives to pursue these consolidated and efficiently packaged space activities.

---

[11] Hosted Payload Alliance; available online at: http://www.hostedpayloadsummit.com/.

[12] Nav Canada; available online at: http://www.navcanada.ca/.

[13] Online at: http://www.iridium.com/About/IridiumNEXT/HostedPayloads.aspx.

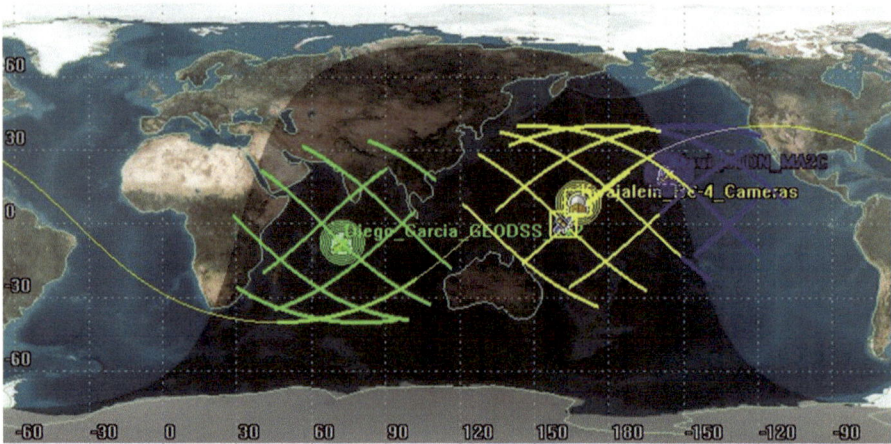

Fig. 3.8 Small satellite orbit designed for greatest observability from designated observation locations

3.2.3 Observability

If an object in orbit cannot be maneuvered, knowing where it is or might be at any point in time is critical. The first consideration is that the object must be discernible either passively by virtue of its own emissions or reflections of background radiation or through active illumination. The degree to which the object's state of motion can be determined or its future state estimated depends on the distribution of observation opportunities and the density of observations acquired during each observation interval.

Observability should be among the principal considerations for the design of the vehicle and the choice of orbit. As an example, consider a single small satellite for which there are sufficient maximum optical observation opportunities. Assume that mission requirements allow any reasonable altitude or inclination. The task is to find an orbit for which there is the most time for cumulative observation by a small set of ground-based sensors.

Safe operation generally requires some compromise in mission capability. For our single satellite to see most of Earth over time, the inclination and apogee should be as high as reasonably possible – taking into account the location of ground observation sites. For example, if one wishes to monitor synoptic energy balance, there would only be brief opportunities for the designated sensors to gather data for orbit estimation. The bold lines in Fig. 3.8 show where the satellites would be visible to the ground observation sites.

3.2.4 Communication and Controllability

A small fraction of satellites intentionally have no communication ability. These are, for example, small satellites whose ballistic coefficients are known precisely and

Fig. 3.9 Nano satellite communication module and antenna (Courtesy of ISIS Cube-sat Solutions)

whose surfaces are appropriately faceted and reflective to assure strong returns from passive or active illumination. They are mostly used to calibrate space surveillance sensors or to characterize atmospheric dynamics, since drag may dominate changes in their trajectories, and those changes can be attributed to changes in density.

All other small satellites must be able at the very least to downlink data, if not respond to commands from the ground. These communication links enable ranging at least and perhaps angular resolution sufficient for reasonable orbit determination. However, observations of this nature are gathered over extremely short arcs and are often conducted with small antennas with poor angular resolution. Gathering and processing sufficient information to determine orbits may require several passes, and there can be gaps between observations that are long enough for orbits to change materially due to environmental variability as a result of intensive solar radiation or other factors (Fig. 3.9).

3.2.5 Maneuverability

The maneuverability of small satellites depends on the key variables in the rocket equation. The ability to change a satellite's velocity depends on how much propellant is available and how much of the initial mass of the satellite is propellant. Electromagnetic thrusters have specific impulses of thousands of seconds of thrust at very low levels. If 90 % of a nano satellite's mass were propellant, total delta V could be about 1 km/s. This, however, is still a small fraction of low Earth orbit velocity. An inclination change of one degree would require a few hundred meters per second of velocity change. If only 10 % of a nano sat mass were propellant, only a few modest maneuvers would consume the entire capability. Independent of overhead mass and power requirements associated with thruster maneuverability, the bottom line is that one cannot expect much collision avoidance maneuverability from a nano sat, even if it is of an eight unit size.

The limited ability of a small satellite to maneuver is still better than a totally uncontrolled object in orbit. A small satellite may also exploit aerodynamics even in the sparse atmosphere of low Earth orbit. The degree of maneuver depends on the

architecture of control surfaces exposed to the environment and the physical characteristics of the environment. A comprehensive review of satellite aerodynamics is available from several sources, including the widely available Wiley *Aerospace Engineering
Encyclopedia*.[14] Aerodynamic attitude or orbit control is efficient in that it relies on the
upper atmosphere as an energy source, but these techniques are generally unreliable,
particularly for collision avoidance purposes. It is impossible to develop avoidance
maneuvers in advance with high probability because satellite trajectories cannot be
estimated with actionable precision more than a few tens of hours in advance, particularly as a result of the drag-dominated low Earth orbits in which they fly.

According to the reports of the International Network of 50 Double and Triple
Cube-sats,[15] aerodynamic forces in the extremely rarefied low Earth orbit regime are
very difficult to estimate. Momentum transfer depends on the physical characteristics of satellite surfaces, which change as the satellite is exposed to the environment.
There have been notable successes, such as the descent of Curiosity to Mars, and
notable failures, such as the Beagle Mars mission.

Propulsive maneuvering capability, when and if available, is thus more suitable.
Propulsion requires stored energy and mass. Cube-sat architecture and missions do
not allow much mass to be allocated to stored propellants. Chemical propulsion is
generally not a viable option for maneuvering. For a variety of reasons related to
minimal mass and low-level but quite sustained thrust, electromagnetic propulsion
is best. Stored high pressure gas or fluids that can be catalyzed to a high pressured
gaseous state with adequate safety and control may also provide suitable propulsion
alternatives.

All of these possibilities are practical for long term, modest orbit or attitude
adjustment, but they seem unsuitable or unreliable for relatively short-notice collision avoidance. Small satellites on a collision course with other small satellites have
no avoidance alternatives. Since desirable missions all favor the same orbit regimes,
collisions among small satellites should not be discounted. Conjunction management between small satellites and larger satellites that can maneuver enough to
avoid catastrophe becomes the sole responsibility of the larger satellite, which
requires more energy to adjust its orbit than the small satellite would.

3.2.6 Assessing Technology Gains Related to Small Satellite Performance

Generally ongoing technology gains continue apace in all aspects of small satellite
design and development. Contemporary power systems are able to generate and
store more power with less mass and volume. Phased-array antennas and deployable

---

[14] David Finkleman, "Atmospheric Interactions with Spacecraft", Wiley Encyclopedia of Aerospace
Engineering, 2010.

[15] http://ec.europa.eu/enterprise/policies/space/files/qb50_en.pdf.

mesh antennas with phased-array multi-beam feed systems are becoming more economical and capable.

The biggest gains have come from turbo-coding technology, which allows these new and efficient encoding systems to transmit more information per bit transmitted. In general, miniaturized electronics and optics and improved processing and encoding techniques have allowed the biggest gains in small satellite technology. Since satellites today are essentially digital processing units in the sky with specialized software that defines what mission they can carry out, such progress is to be expected. In short, gains in the field of computer technology and computer science programming can generally be transferred to the field of artificial satellites. Thus, parallel gains largely come in the rapidly evolving fields of both computer systems and satellites.

There is yet another area of new technology development that is particularly relevant to the policy and regulatory issues for small satellites that also needs to be given particular attention. This is the area of technology that would allow small satellites to pose less of an issue or concern with regard to the increasingly troublesome issue of orbital debris and de-orbit of small satellites in low Earth orbits that are today becoming more and more congested.

3.3   De-orbit Capabilities for Small Satellites

Active debris removal is imperative since, even if no new space objects are launched, the number of objects already in orbit would create so much more space debris that the use of space might not be sustainable on a business as usual basis. Therefore, various technical means and de-orbiting capabilities have to be developed to support active debris removal.[16]

There are a number of ways to address the orbital debris problem as it relates to small satellites, but in a broader sense, these innovations fall into one of two categories: (i) ways to help de-orbit small satellites more efficiently; and, (ii) ways to repackage small payloads into larger and more efficient systems so that there are fewer of them going into space or, alternatively, they can be de-orbited more effectively as part of larger system.

Incentives to create separate free-flyer small satellite missions remain. This means that the problem of de-orbiting of small satellites at end-of-life remains a very real challenge.

Currently, there is great interest in the development of new technology to assist with de-orbit of these spacecraft. There are several concepts about how this might be done for small satellites and especially for nano satellites with no thrusters or active mechanism to initiate de-orbit. These include inflatable and reflective

---

[16] "Active Debris Removal – An Essential Mechanism for Ensuring the Safety and Sustainability of Outer Space: A Report of the International Interdisciplinary Congress on Space Debris Remediation and On-Orbit Satellite Servicing," UN Document: A/AC.105/C.1/2012/CRP.16 of 27 January 2012.

balloon-like membranes,[17] inflatable tube structures with thin membranes (known as ITMs),[18] solar sail systems, and tether systems. The idea is that all of these low-mass systems could be either inflated or deployed at the end of life of a small satellite to accelerate its descent from low Earth orbit back to the ground. Many of these de-orbit systems are student projects at research universities. However, NASA's Fastrac Satellite included a 4.0-kg experiment called the NANO-SAIL-D2 that was designed to be deployed from the FASTRAC satellite pictured earlier. When fully deployed, this thin membrane extends up to 100 square feet, or about 9 square meters. Since this solar sail was itself a 3U nano satellite it is clear that such a solar sail to assist with de-orbit could be deployed only for bigger small satellites. This experiment was not a total success in that it was planned to deploy the NANO-SAIL D2 2 weeks after the launch of the FASTRAC satellite on December 3, 2010, but deployment was not achieved as scheduled. Then, for reasons that have not been entirely explained, the NANO-SAIL D2 self-deployed some 6 weeks later on January 17, 2011.[19] In addition, there are planned experiments with tether systems that could aid small satellite de-orbit.

The design of systems that could allow effective and low cost de-orbit of small satellites remains a well-focused area of research. And passive systems to accelerate de-orbit are not only being developed but will likely soon be offered on a commercial basis. Not all de-orbit systems for small and nano-satellites today are entirely based on passive systems. The Surrey Satellite Technology, Ltd., group has developed a micro-thruster system which they are now testing after their successful launch of their latest nanosatellite in February 2013. SSTL and SSC have several upcoming nano- and micro-satellites that will demonstrate the use of deployable sails first to reduce orbital lifetime by increasing drag, and then later to demonstrate the active capture of space debris and de-orbiting by the use of a drag sail.

The STRaND-1 3 U Cube-sat, also shown earlier in Fig. 3.7 above, contains an active micro-thruster system to assist with de-orbit. The active de-orbit system flying on this remarkable nano-satellite is about the size of a loaf of bread and was designed and built by volunteers in a span of only about 3 months. In an apparent reference to the "Star Trek" sci-fi series, this active de-orbit mechanism is called WARP DRIVE. In this instance, however, the name stands for Water Alcohol Resisto-jet Propulsion Deorbit Re-entry Velocity Experiment, and it consists of eight micro-pulse plasma thrusters.[20]

---

[17] C. Lucking, A Passive High Altitude Deorbiting Strategy Advanced Space Concepts Laboratory, University of Strathclyde; available online at: https://pure.strath.ac.uk/portal/files/5443747/Heiligers_J_Colombo_C_McInnes_CR_Pure_A_passive_high_altitude_deorbiting_strategy_08_Aug_2011.pdf.

[18] Y. Miyazaki et al., "A Deployable Membrane Structure for De-Orbiting a Nano-satellites IAC-07-B4.5.08 (2007); available online at: http://www.iafastro.net/iac/archive/browse/IAC-07/B4/5/7019/.

[19] NASA to Attempt Historic Solar Sail Deployment; available online at: http://science.nasa.gov/science-news/science-at-nasa/2008/26jun_nanosaild/.

[20] WARP DRIVE to be tested on Surrey Space Technology Ltd. STRaND-1 nano-satellite; available online at: http://www.sstl.co.uk/Missions/STRaND-1--Launched-2013/STRaND-1/STRaND-1-FAQs.

Fig. 3.10 Artist representation of "Kickstarter" Plasma Thruster with cube-sat (Courtesy of NASA)

Researchers at the University of Michigan's Aerospace Engineering department are currently working in collaboration with several NASA research centers and private industry as part of what is known as the Kickstarter campaign (Fig. 3.10). This initiative is seeking to develop the Cube-sat Ambipolar Thruster (CAT), a new type of plasma propulsion system. It is hoped that this plasma thruster system would be able to propel cube-sats at low thrust levels in gradually increasing spiral orbits so that they would be able to escape Earth's gravity and go into deep space. Researchers are claiming that they can accomplish this at very low cost.

On one hand, such a system could help remove debris from low Earth orbit. However, long-term spiral orbit deployment from low Earth orbit to deep space could create a risk of collision with orbital debris during the orbit-raising exercise. Thus, a careful risk assessment of this approach is clearly needed.[21]

Clearly, the design and deployment of satellites, including small satellites, involves a great deal of technology and operational expertise. Yet, this is only half of the process. In the next few chapters, key concepts relating to the deployment of satellites in terms of legal, regulatory, licensing, registration, and frequency management issues and processes will be presented. This will be followed by a discussion of the problem of orbital debris, especially in the context of small satellites and then the regulatory processes that have sought to address this issue as well as the responsibility and liability provisions that apply to space objects, especially small satellites.

---

[21] University of Michigan Kickstarter Campaign to develop Plasma Thruster for Cube-sat Missions; available online at: http://www.kickstarter.com/projects/597141632/cat-a-thruster-for-interplanetary-cube-sats.

Chapter 4
The Global Legal Guidelines Governing Satellite Deployment

4.1 Introduction

All small satellites, irrespective of their size, weight, and scope of missions, are space objects that are governed by currently applicable international legal guidelines. Secondly, the launching and operation of such satellites constitute space activities in the form of exploration and use of outer space. Thus, they are also subject to such guidelines. These guidelines have been established primarily through the United Nations. It consists of five major international treaties negotiated through the U. N. Committee on Peaceful Uses of Outer Space.[1] In addition to these treaties,

---

[1] The Treaty on Principles Governing the Activities of States in the Exploration and Use of Outer Space, including the Moon and Other Celestial Bodies (the "Outer Space Treaty"), adopted by the General Assembly in its resolution 2222 (XXI), opened for signature on 27 January 1967, entered into force on 10 October 1967, there are 102 ratifications and 26 signatures (as of 1 April 2013); the Agreement on the Rescue of Astronauts, the Return of Astronauts and the Return of Objects Launched into Outer Space (the "Rescue Agreement"), adopted by the General Assembly in its resolution 2345 (XXII), opened for signature on 22 April 1968, entered into force on 3 December 1968, there are 92 ratifications, 24 signatures, and 2 acceptance of rights and obligations (as of 1 April 2013); the Convention on International Liability for Damage Caused by Space Objects (the "Liability Convention"), adopted by the General Assembly in its resolution 2777 (XXVI), opened for signature on 29 March 1972, entered into force on 1 September 1972, there are 89 ratifications, 22 signatures, and 3 acceptances of rights and obligations (as of 1 April 2013); the Convention on Registration of Objects Launched into Outer Space (the "Registration Convention"), adopted by the General Assembly in its resolution 3235 (XXIX), opened for signature on 14 January 1975, entered into force on 15 September 1976, there are 61 ratifications, 4 signatures, and 2 acceptances of rights and obligations (as of 1 April 2013); and the Agreement Governing the Activities of States on the Moon and Other Celestial Bodies (the "Moon Agreement"), adopted by the General Assembly in its resolution 34.68, opened for signature on 18 December 1979, entered into force on 11 July 1984, there are 15 ratifications and 4 signatures (as of 1 April 2013).

R.S. Jakhu and J.N. Pelton, *Small Satellites and Their Regulation*, SpringerBriefs in Space Development, DOI 10.1007/978-1-4614-9423-2_4, © Springer New York 2014

there are several U. N. Regulations and Guidelines,[2] principles, and rules of general international law, and some other international agreements[3] that are directly applicable to small satellites.

4.2 Rights of Space Use by Large and Small Satellites

It can be said that, in general, all international rights and obligations of the states with respect to big satellites are equally relevant for the conduct of space activities involving the use of small satellites. Here is a brief list of such rights and obligations:

- All states and their non-governmental entities (i.e., private citizens, companies, universities) are entitled to freely explore and use outer space without discrimination of any kind, on a basis of equality and in accordance with international law.[4]
- States and their non-governmental entities are prohibited appropriating outer space by claim of sovereignty, by means of use or occupation, or by any other means.[5]

---

[2] They are: (a) the UN *Principles Governing the Use by States of Artificial Earth Satellites for International Direct Television Broadcasting*, adopted by the UN General Assembly by 107 votes to 13, with 13 abstentions, on 10 December 1982 (under a General Assembly resolution 37/92: voting results are reproduced from UN document A/37/PV.100 of 17 December 1982); (b) the 1963 UN *Declaration of Legal Principles Governing the Activities of States in the Exploration and Use of Outer Space*, adopted by the UN General Assembly (under General Assembly resolution 1962(XVIII) on 13 December 1963); (c) the 1986 UN *Principles Relating to Remote Sensing of the Earth from Outer Space*, adopted by the UN General Assembly without vote (under General Assembly resolution 41/65 on 3 December 1986); (d) the 1992 UN *Principles Relevant to the Use of Nuclear Power Sources In Outer Space*, adopted by the UN General Assembly without vote (under General Assembly resolution 47/68 on 14 December 1992); (e) the 1996 *Declaration on International Cooperation in the Exploration and Use of Outer Space for the Benefit and in the Interest of All States, Taking into Particular Account the Needs of Developing Countries*, adopted by the UN General Assembly without vote (under UN General Assembly Resolution A/RES/51/122 on 13 December 1996); and (f) the 2007 *Space Debris Mitigation Guidelines of the Committee on the Peaceful Uses of Outer Space,* Official Records of the General Assembly, Sixty-second Session, Supplement No. 20(A/62/20), paras. 117 and 118 and annex. The UN General Assembly in its Resolution endorsed the Space Debris Mitigation Guidelines of the Committee on the Peaceful Uses of Outer Space in 2007. See: United Nations General Assembly, Sixty-second session, Agenda item 31, Document A/RES/62/217 (10 January 2008), paragraph 26.

[3] The most important of such agreements are the 1945 Charter of the United Nations and the Constitution and the Constitution and Convention of the International Telecommunication Union, 1994 (as amended in 2012; hereinafter referred to as ITU Constitution) and ITU Radio Regulations, Edition of 2012 (as amended; hereinafter referred to as ITU Radio Regulations). Currently, there are 192 States Parties these instruments.

[4] Article I, the 1967 Outer Space Treaty.

[5] Article II, the 1967 Outer Space Treaty.

- There is a prohibition on the placement in orbit around Earth any objects carrying nuclear weapons or any other kinds of weapons of mass destruction.[6]
- States are internationally responsible for their national (public and private) activities in outer space.[7]
- States are internationally responsible for assuring that their national (public and private) activities are carried out in conformity with the provisions set forth in the Outer Space Treaty.[8]
- The activities of non-governmental entities (i.e., private citizens, companies, universities) in outer space must be carried out under "authorization and continuing supervision by the appropriate state party to the Outer Space Treaty."[9]
- Each launching state (not its citizens nor private companies) is internationally liable for damage to another state party to the Outer Space Treaty (or to its citizens or private companies) caused by its space object or its component parts. The term 'launching state' refers to a state that launches or procures the launching of a space object and a state from whose territory or facility a space object is launched.[10]
- States are to be guided by the principle of cooperation and mutual assistance and are encouraged to conduct all their activities in outer space with due regard to the corresponding interests of all other states and their public and private entities.[11]
- Each launching state is obliged to register its (and those belonging to its citizens or private companies or universities) space objects in its national register.[12]
- Each launching state must also inform the U. N. Secretary-General about its launched space objects so that international registration can be carried out.[13]
- The ownership of objects launched into outer space is not affected by their presence in outer space or on a celestial body or by their return to Earth.[14]
- A state on whose registry an object launched into outer space is carried must retain jurisdiction and control over such object, and over any personnel thereof, while in outer space or on a celestial body.[15]
- Each state must carry out its space activities (and ensure that those belonging to its citizens or private companies or universities are carried out) for "peaceful purposes,"[16] which includes military but excludes "aggressive" purposes. States are entitled to use space for self-defense purposes as specified under Article 51 of the U. N. Charter.

---

[6] Article IV, the 1967 Outer Space Treaty.

[7] Article VI, the 1967 Outer Space Treaty.

[8] Article VI, the 1967 Outer Space Treaty.

[9] Article VI, the 1967 Outer Space Treaty.

[10] Article VII, the 1967 Outer Space Treaty; the 1972 Liability Convention.

[11] Article IX, the 1967 Outer Space Treaty.

[12] Article II (1), Registration Convention.

[13] Article IV, Registration Convention.

[14] Article VIII, the 1967 Outer Space Treaty.

[15] Article VIII, the 1967 Outer Space Treaty.

[16] Preamble, the 1967 Outer Space Treaty.

- All states are obliged to establish and operate their radio stations, including those onboard small satellites, in such a manner as not to cause harmful interference to the radio services of others that are operated in accordance with the provisions of the ITU Radio Regulations.[17]
- States are required to take all practicable steps to prevent the operation of small satellites and installations of all kinds from causing harmful interference to the radio services of others.[18]
- Each state is obliged to require its citizens or private companies to respect these obligations.[19] Therefore, no transmitting radio system (station) can be established or operated by a private person or by any enterprise without a license issued in an appropriate form and in conformity with the provisions of the ITU Radio Regulations by or on behalf of the government of the country to which the station in question is subject.[20]

International law is essentially applicable to states. The rights relating to freedom of exploration and use of outer space, as guaranteed for states under international agreements and treaties, are exercised by states, and states exclusively remain responsible and liable for the activities of their citizens or private companies. In the exercise of their rights, states may allow their respective citizens or private companies or academic institutions access to space, and may impose restrictions on such use as they consider necessary. States also pass on their international obligations to their respective citizens or private companies. This is essentially done under national legal and regulatory regimes and policies. Each state adopts its national laws and regulations, the nature, scope, and timing of which is essentially determined by its politico-economic policies and priorities.

4.3   National Laws Relating to Usage of Space

Almost all the countries that are involved in space exploration and use have some form of national laws, regulations and/or administrative directives and policies under which space activities are carried on. These laws could be classified as follows:

(a) Specific laws regulating general national space activities and domestically implementing obligations under international space treaties such as the Outer Space Treaty, the Registration Convention, the Liability Convention, the

---

[17] Article 45, ITU Constitution.

[18] Article 45, ITU Constitution.

[19] Article 45, ITU Constitution.

[20] Article 18 (1), ITU Radio Regulations.

Swedish Space Activities Act,[21] the U.K. Outer Space Act,[22] the Space Act of South Arica,[23] the Australian Space Act,[24] and so on.

(b) Minor amendments to the current laws or supplementing existing laws to extend their scope to cover space activities. This is the most common practice; an example is minor changes in the Radiocommunications Act of Canada[25] to cover the licensing of telecommunications satellites.

(c) Laws establishing organizations to carry out space research or other space activities, e.g., JAXA (Japan),[26] NASA (USA),[27] CSA (Canada),[28] National Space Agency Act (South Africa), etc.[29]

(d) New legislation specifically adopted to regulate a particular space activity or space application; for example, in the United States separate acts regulate launch services,[30] remote-sensing activities,[31] and satellite telecommunication services.[32] In Canada, the Remote-Sensing Act governs remote-sensing activities of both public and private sector.[33]

(e) Various administrative directives and/or policies that specifically apply to some aspects of space activities that are carried out primarily by governments, such

---

[21] Act on Space Activities, (1982:963); available online at: http://www.oosa.unvienna.org/oosa/en/SpaceLaw/national/sweden/act_on_space_activities_1982E.html.

[22] Outer Space Act, (1986 Chapter 38); available online at: http://www.legislation.gov.uk/ukpga/1986/38/introduction.

[23] Space Affairs Act, (Statutes of the Republic of South Africa - Trade and Industry No. 84 of 1993; assented to 23 June 1993; date of commencement: 6 September, 1993); available online at: http://www.oosa.unvienna.org/oosa/en/SpaceLaw/national/south_africa/space_affairs_act_1993E.html. Space Affairs Amendment Act, (No. 1530. 6 October 1995); available online at: http://www.oosa.unvienna.org/oosa/en/SpaceLaw/national/south_africa/space_affairs_amendment_act_1995E.html.

[24] Space Activities Act, (Act No. 123 of 1998 as amended); available on line at: http://www.comlaw.gov.au/Details/C2010C00193.

[25] Radiocommunication Act, (R.S.C., 1985, c. R-2,); available online at: http://laws-lois.justice.gc.ca/eng/acts/R-2/page-1.html.

[26] Law Concerning Japan Aerospace Exploration Agency, (Law Number 161 of 13th December 2002); available online at: http://www.jaxa.jp/about/law/law_e.pdf.

[27] The National Aeronautics and Space Act, (as amended, Pub. L. No. 111–314; 124 Stat. 3328, Dec. 18, 2010), now codified in 51 U.S.C. § 20113(a); available online at: http://uscode.house.gov/download/pls/Title_51.txt.

[28] Canadian Space Agency Act, (S.C. 1990, c. 13, assented to 1990-05-10); available online at: http://laws.justice.gc.ca/eng/acts/C-23.2/page-1.html.

[29] South African National Space Agency Act, (No. 36 of 2008); available online at: http://www.oosa.unvienna.org/pdf/spacelaw/national/safrica/Act36-2008.pdf.

[30] Commercial Space Launch Activities Act of 1984 (as amended), now codified in 51 USC Chapter 509, available online at: http://www.law.cornell.edu/uscode/text/51/subtitle-V/chapter-509.

[31] Land Remote Sensing Policy Act of 1992, (now codified in 51 USC Chapter 601); available online at: http://www.nasa.gov/offices/ogc/commercial/15uscchap82.html.

[32] Communications Act of 1934, (47 USC Chapter 5); available online at: http://www.law.cornell.edu/uscode/text/47/chapter-5; and Satellite Communications Regulations, (47 Code of Federal Regulations Part 25); available online at: http://www.law.cornell.edu/cfr/text/47/25.

[33] Remote Sensing Space Systems Act, (S.C. 2005, c. 45, assented to 2005-11-25); available online at: http://laws-lois.justice.gc.ca/eng/acts/R-5.4/page-1.html.

as the 2011 Remote-Sensing Data Policy of India[34] and the Norms, Guidelines and Procedures for Implementation of the Policy Framework for Satellite Communications in India.[35]

The United States is a world leader in national regulation of space activities. Recently, a few other states have started adopting some forms of national space laws and regulations.[36] However, a large majority of states (including spacefaring nations) do not have effective national laws and regulations to govern all the various space activities, including the launch and use of small satellites.

---

[34] Available online at: http://www.isro.org/news/pdf/RSDP-2011.pdf.

[35] Available online at: http://www.isro.org/news/pdf/SATCOM-norms.pdf.

[36] Ram S. Jakhu, (Ed.), *National Regulation of Space Activities*, 2010, published by Springer Publishing House, the Netherlands.

Chapter 5
Licensing, Registration, and Frequency Use Regulation

5.1 Launch and Payload Licenses

The most common and important tool for nationally regulating space activities is the requirement of a license (authorization) from a designated governmental entity. A national license creates a legal nexus between the issuing state and the licensee. Such a license determines the rights and obligations of the licensee; is normally issued under certain conditions and terms; and could be terminated if the licensee acts contrary to the provisions of the applicable law, regulations, and the conditions and terms of the license.

State practice shows a variety of licensing requirements and processes related to small satellites. For example, in Canada the Department of Foreign Affairs and International Trade (DFAIT) has issued several licenses for remote-sensing systems under the Canada Remote-Sensing Space System Act.[1] However, NEOSSat (Near Earth Object Surveillance Satellite), a Canadian small satellite weighing only 65 kg, which was recently launched into an 800-km orbit using an Indian Polar Satellite Launch Vehicle (PSLV), was not licensed by the DFAIT since it is a "space telescope dedicated to detecting and tracking asteroids and satellites"[2] and not a remote-sensing satellite as defined in the Act.[3]

Perhaps another reason why NEOSSat was not licensed by DFAIT was because it was exempted from licensing by virtue of an order in council issued by Her Majesty. Such exemptions are provided for in the act where governmental missions are carried out. The main implication of this legal lacuna in licensing jurisdiction on the part of

---

[1] Remote Sensing Space Systems; available online at: http://www.international.gc.ca/arms-armes/non_nuclear-non_nucleaire/remote_sensing-teledetection.aspx?lang=eng.

[2] NEOSSAT: Canada's Sentinel in the Sky; available online at: http://www.asc-csa.gc.ca/eng/satellites/neossat/.

[3] Section 2 of the Remote Sensing Space Systems Act, *supra* note 71, defines "*remote sensing satellite*" as "a satellite that is capable of sensing the surface of the Earth through the use of electromagnetic waves."

R.S. Jakhu and J.N. Pelton, *Small Satellites and Their Regulation*, SpringerBriefs in Space Development, DOI 10.1007/978-1-4614-9423-2_5, © Springer New York 2014

the DFAIT is that though NEOSSat is an unlicensed small satellite, Canada is its launching state under international space law since the satellite has been funded and is being operated jointly by two governmental agencies – the Defense Research Development Canada and the Canadian Space Agency. Consequently, the government of Canada can be held liable if any damage is caused by NEOSSat.

Sometimes, countries lack an appropriate national regulatory system to issue licenses for a particular space activity that could be carried out with big or small satellites. It is interesting to note that small satellites are being designed for on-orbit satellite servicing (OOS) activities. For example, ATK's small satellite bus (A500) has already been used "for the revolutionary DARPA Phoenix mission to conduct on-orbit satellite servicing and repurposing."[4] The Canadian company McDonald Dettwiler and Associates Ltd. (MDA) has developed robotic capability that could be used for on-orbit satellite servicing activities. Intelsat and MDA had agreed to conduct such operations for refueling missions.[5] Although the deal failed mainly due to lack of significant business prospects, it would have been difficult for MDA to procure a license in Canada since the currently applicable Canadian regulatory regime makes no provision for such activities.

In Canada, pursuant to Section 5 of the Canadian Remote-Sensing Space Systems Act, no person is permitted to operate a remote-sensing satellite system without first procuring a license from the Minister of Foreign Affairs. Before a license is issued, the minister is required to consider national security, the defense of Canada, the safety of Canadian forces, and Canada's international relations and obligations.

Such licenses are issued subject to several conditions, including the following: (a) that the licensee keep control of the licensed system; (b) that the licensee not permit any other person to carry on a controlled activity in the operation of the system except in accordance with the license; (c) that raw data and remote-sensing products from the system about the territory of any country (but not including data or products that have been enhanced or to which some value has been added) be made available to the government of that country within a reasonable time and on reasonable terms, but subject to any license conditions that the minister considers appropriate; (d) that the licensee keep control of raw data and remote-sensing products from the system; and, (e) that the licensee must dispose of its satellite at the end

---

[4] "ATK Introduces Expanded Product Line of Small Satellite Spacecraft Platforms," July 30, 2012; available online at: http://www.atk.com/news-releases/atk-introduces-expanded-product-line-of-small-satellite-spacecraft-platforms/

"MDA signs satellite services deal with Intelsat," Mar. 15, 2011; available online at: http://www.theglobeandmail.com/globe-investor/mda-signs-satellite-services-deal-with-intelsat/article578531/

"ATK Introduces Expanded Product Line of Small Satellite Spacecraft Platforms," July 30, 2012; available online at: http://www.atk.com/news-releases/atk-introduces-expanded-product-line-of-small-satellite-spacecraft-platforms/.

[5] "MDA signs satellite services deal with Intelsat," Mar. 15, 2011; available online at: http://www.theglobeandmail.com/globe-investor/mda-signs-satellite-services-deal-with-intelsat/article578531/.

of its life in a manner described in its disposal plan that was approved by the minister before the issuance of the license.[6]

In the United States, a license is required in order to operate any private space remote-sensing system by any person who is subject to the jurisdiction or control of the United States.[7] The license is issued by the Secretary of Commerce, who has delegated this authority to the National Oceanic and Atmospheric Administration (NOAA). To procure a license, the applicant (and the licensee) must comply with several broad and onerous requirements, including the duty to respect international obligations and national security concerns of the United States as well as to maintain operational control of the satellite from within the United States; limitations on the collection and dissemination of data; and disposal of the satellite at end-of-life in a manner approved by NOAA.

The current U. S. law is applicable to all individuals, academic institutions, and private companies that are planning to launch big or small remote-sensing satellites, "even though [cube-sats] often don't raise national security or foreign policy concerns."[8] NOAA recognizes that it will be difficult to strictly apply the requirements of the law to small satellites, particularly, since "[d]esign of some cube-sat systems makes it impossible to comply with standard licensing conditions, e.g., limitation of imaging operations when required by national security concerns."[9] Therefore, one can expect an eventual lessening of the licensing burden on small satellites in the United States.

Glenn Tallia (Chief, Weather Satellites and Research Section, NOAA General Counsel) believes that, in addition to the development of guidelines and codes of conduct, there is the need to change current remote-sensing regulations to provide NOAA with the "discretion to determine that certain cube-sats that propose to image Earth do not require a license."[10] In addition, there might also be changes, according to Tahara Dawkins (Director of NOAA's Commercial Remote-Sensing Regulatory Affairs office) that could include "the possible reduction of resolution restrictions, to help U. S. commercial data providers maintain their competitive advantage and retain market leadership while continuing to take into account U. S. national security concerns."[11]

---

[6] Ram S. Jakhu, Catherine Doldirina and Yaw Otu Mankata Nyampong, *Review Of Canada's Remote Sensing Space Systems Act Of 2005*, 2012, Annals of Air and Space Law, p. 399.

[7] National and Commercial Space Policy Act of 2010 (formerly the 1992 Land Remote Sensing Policy Act), 51 USC § 60122.

[8] Glenn Tallia (Chief, Weather Satellites and Research Section, NOAA General Counsel), "NOAA's Licensing of Cube-sats as Private Remote Sensing Space Systems under the National and Commercial Space Policy Act", January 20, 2012; available online at: http://www.americanbar. org/content/dam/aba/administrative/science_technology/1_20_12_licensing.authcheckdam.pdf.

[9] *Ibid.*

[10] *Ibid.*

[11] Harrison Donnelly, "Remote Sensing Regulator, GIF 2013 Volume: 11 Issue: 1 (February); available online at: http://www.kmimediagroup.com/mgt-home/466-gif-2013-volume-11-issue-1-february/6353-remote-sensing-regulator.html.

These proposals certainly look positive and should encourage the development and launch of small satellites. However, this is a two-edged sword. The more streamlined process to launch can be seen as a positive step, but if there are many more small satellite launches that contribute to the increase of orbital debris, this must be considered a negative result.

In particular, it must be kept in mind that it is the capability of a satellite rather than its size or weight that determines the scope and nature of the licensing burden. As the technology develops, satellites that are small in size have started carrying out more advanced and sophisticated remote-sensing activities.[12] According to Siegfried Janson, a senior scientist at the Aerospace Corporation in El Segundo, California, "[T]oday's small satellites are as capable, or more so, than their larger cousins were 25 years ago. This trend should continue over the next 25 years."[13]

There is a growing trend not only towards the construction of small satellites but also towards the development of small launch vehicles. Such reusable launchers will be routinely placing small satellites, including personal satellites,[14] into orbit and possibly at a rate of four or more times a day. In collaboration with Surrey Satellite Technology, Virgin Galactic is developing its "'LauncherOne' program, an unmanned rocket that will be air-launched by SpaceShipTwo's carrier aircraft, WhiteKnightTwo, which will be capable of delivering as much as 225 kg to low Earth orbit."[15] Licenses for such small launches with small payloads will pose challenges for regulatory authorities. In the United States, all launch services are regulated under the 1984 Commercial Space Launch Services Act (as amended).[16] A license is required for expandable and reusable launch vehicles for private launch

---

[12] For details see, "Small Satellites Redefine Earth Observation"; available online at http://www.sst-us.com/blog/march-2013/small-satellites-redefine-earth-observation; Janet French, "Small satellite, big dreams," The Star Phoenix, 24 August 2011; available online at http://www2.canada.com/saskatoonstarphoenix/news/story.html?id=8a8dbcf9-37ee-4013-b208-f8a3dc7d2cdb (last accessed: 2 May 2013); Andrew Cawthorne, David Purll, and Stuart Eves, "Very High Resolution Imaging Using Small Satellites," a paper presented at 6th Responsive Space Conference, April 28–May 1, 2008, Los Angeles, CA; available online at: http://www.responsivespace.com/Papers/RS6/SESSIONS/SESSION%20III/4007_CAWTHORNE/4007P.pdf; "SwRI Building Eight NASA Nano-satellites to Help Predict Extreme Weather Events on Earth," Jun 25, 2012; available online at: http://www.spacedaily.com/reports/SwRI_Building_Eight_NASA_Nano-satellites_to_Help_Predict_Extreme_Weather_Events_on_Earth_999.html.

[13] Leonard David, "Small Satellites Finding Bigger Roles as Acceptance Grows, Aug. 29, 2011; available online at: http://www.spacenews.com/article/small-satellites-finding-bigger-roles-acceptance-grows#.UZpI3rXVB8E.

[14] A personal satellite can be cracker-size satellite, that's "cheap enough for average people to build and fly their own satellite": Caleb Garling, Entrepreneur working on personal satellites, December 26, 2012; available online at: http://www.chron.com/business/technology/article/Personal-satellites-that-fly-into-space-4146595.php.

[15] "SST US collaborates with Virgin Galactic to offer radically cheaper options for small satellites," July 12, 2012; available online at: http://www.spacedaily.com/reports/SST_US_collaborates_with_Virgin_Galactic_to_offer_radically_cheaper_options_for_small_satellites_999.html.

[16] Commercial Space Launch Activities, 51 USC 509.

operations (a) within U. S. territories by anyone, (b) outside the United States by U. S. citizens for citizens of the United States, (c) outside the United States and outside the territory of a foreign country unless there is an agreement between the United States and the government of the foreign country providing that the government of the foreign country has jurisdiction, and (d) for a citizen of the United States in the territory of a foreign country if there is an agreement between the United States and the government of the foreign country providing that the United States has jurisdiction.[17]

The launch licenses are issued by the office of the Secretary of Transportation, which has delegated its authority to the Office of Commercial Space Transportation (AST) within the Federal Aviation Administration (FAA). A license is issued after a thorough safety and mission review has been conducted that involves consultations with and decisions by various agencies of the U. S. government. Before issuing a launch license, the AST must insure that an appropriate license has been obtained from the Federal Communications Commission (FCC) with respect to communication satellites and the Department of Commerce with respect to remote-sensing satellites. In addition, AST must ensure the protection of the 'national security interest' of the United States in consultation with the Department of Defense, and the 'foreign policy interests or obligations of the United States' in consultation with the Department of State.

A license may then be issued subject to various conditions, including (a) the requirement of a minimum amount of third-party liability insurance, the amount of which is determined on basis of the maximum probable loss; (b) strict adherence to safety regulations; (c) requirements concerning pre-launch records and notifications including those pertaining to FAA airspace restrictions; and (d) compliance with federal inspection, verification, and enforcement requirements.

Such requirements and the process for procuring launch licenses will prove cumbersome for routine launches of small satellites by small launch vehicles such as LauncherOne. However, U. S. law permits the Secretary of Transportation to waive any requirement, including the requirement to obtain a license, for an individual applicant if the Secretary decides that the waiver is in the public interest and will not jeopardize the public health and safety, safety of property, and national security and foreign policy interests of the United States.[18] This authority could be used to waive the requirement of licenses for some small satellites.

India's Polar Satellite Launch Vehicle (PSLV) rockets are increasingly being used as the launch vehicle for small satellites, particularly from foreign countries. India does not have a national law relating to launch licenses. A customer seeking a launch thus only needs to enter into an agreement with Antrix Corporation, the marketing arm of the Indian Space Research Organization (ISRO), which is a part of the Indian Department of Space. However, according to Sridhara Murthi (former

---

[17] 51 USC § 50904.
[18] 51 USC § 50905 (3).

director of Antrix Corporation), all launch requests of "satellites on Indian launch vehicles (other than Indian government-owned) still require authorization for launch from the government of India. The authorizations are provided through the Department of Space and take into account any international treaty obligations, foreign policy interests, and national security considerations."[19] The Department of Space makes its decisions in this regard not on the basis of any launch-related act or regulation but under internal administrative processes and policies of the government of India. As compared to the situation in the United States, the procurement of launch authorizations in India is much simpler and expeditious; thus it is quite conducive to the owners of small satellites, both from India and foreign countries.

South Africa's first spacecraft developed by South Africans – Sunsat – weighed about 64 kg and was launched by an American Delta II rocket in February 1999.[20] This small satellite was not licensed in South Africa. However, the second South African small satellite weighing 80 kg, Sumbandilasat, is a government-commissioned satellite that was launched on September 17, 2009, into a 600-km orbit.[21] A launch license for this satellite was issued on June 2, 2009, to the National Department of Science and Technology by the South African Space Affairs Council under Section 11 of the Space Affairs Act of 1993. ZACUBE-1, South Africa's first cube satellite weighing only 1.3 kg and measuring $10 \times 10 \times 10$ cm, was designed and built mainly by postgraduate students at the Cape Peninsula University of Technology.[22] This spacecraft may be launched in mid-2013[23] and is yet to be licensed.

In Nigeria, the National Space Research and Development Agency (NASRDA) has been established pursuant to the NASRDA Act[24] with the mandate to undertake certain activities pertaining to space.[25] The act also established a Space Council that is empowered, on the recommendation of the agency, to license non-governmental entities and corporate persons desirous of undertaking certain space activities.[26] NigeriaSat-2, a mini satellite weighing about 300 kg and Nigeriasat X, weighing about 100 kg, are wholly owned and operated by the National Space Research and Development Agency (NASRDA). Since these satellites were procured by the NASRDA on behalf of the Nigerian government, no licenses were needed or issued.

[19] K.R. Sridhara Murthi, "ISRO's Launcher Policies and International Services", a paper (IAC-09-E3.3.2) presented at the International Astronautical Congress, 2009, p. 7.

[20] Garth W. Milne, et al., "SUNSAT - Launch and First Six Month's Orbital Performance," a paper presented at 13th Annual AIAA/USU Conference on Small Satellites, 1999; available online at: http://staff.ee.sun.ac.za/whsteyn/Papers/USU99_Sunsat.PDF.

[21] Online at: http://www.ne.jp/asahi/hamradio/je9pel/sumbandi.htm.

[22] "South Africa's First Cube-sat Heads For Space," 2 October 2012; available online at: http://www.sansa.org.za/spacescience/resource-centre/news/83-south-africa-s-first-cube-sat-heads-for-space.

[23] https://directory.eoportal.org/web/eoportal/satellite-missions/v-w-x-y-z/zacube-1

[24] National Space Research and Development Agency Act of 2010, the text of which is available online at: http://reclaimnaija.net/cms/act/2010/national_space_research_and_development_agency_act_2010.pdf.

[25] Ibid, Article 6.

[26] Ibid, Article 9.

There is no doubt that, if these satellites had been procured by any non-governmental entity the issue of licensing would have arisen and would have been addressed.

Like several other nations, in Japan there is no law authorizing the granting of licenses either for launches or for satellite operations. So no governmental license is issued for small satellites.

5.2   Registration

As noted above, under international space law each launching state is obliged to register satellites belonging to its public entities, citizens, or private companies in its national register and also to register such satellites with the U. N. Secretary-General. The requirement of international registration of space objects was adopted pursuant to a belief that a mandatory system of registering space objects would assist in their identification and would contribute to the application of international space law, particularly in determining responsibility and liability in cases of accidents.

There has been good compliance to this by states, as spacefaring nations have been regularly registering their launched space objects in their national registers as well as in the international register maintained by the U. N. Secretary-General under the 1975 Registration Convention or the U. N. General Assembly resolution 1721B (XVI). Although international registration is mandatory under the convention, the number of international registrations of space objects made by states has been declining in recent times. According to a study by the International Law Association (ILA), "[B]efore the 1975 Registration Convention, and under UNGA Resolution 1721B (XVI), 129 objects were launched into outer space in 1972, all of which were registered (0% unregistered objects). In 1990, 165 objects were launched into outer space of which 160 were registered (9% unregistered objects). In 2002, 92 objects were launched into outer space of which 73 were registered (20% unregistered objects). In 2004, 72 objects were launched into outer space of which only 50 were registered (30.5% unregistered objects). Indeed we are going downhill in this regard."[27]

One of the main reasons for this growing reluctance towards international registration is that such registration is required to be carried out as soon as possible, after a satellite has been launched and registered on a national registry.[28] There is, however, no specific time limitation for international registration. States tend to delay or decide not to send the required information to the U. N. Secretary General, particularly regarding those satellites that have been launched by foreign launch vehicles and those that might not remain in orbit for a long time.

---

[27] ILA Space Law Committee, "Legal Aspects of the Privatisation and Commercialisation of Space Activities: Remote Sensing and National Space Legislation", Second Report for the 2006 ILA Toronto Conference, INTRODUCTION by Professor Maureen Williams.

[28] Article IV, Registration Convention.

The American satellite Iridium 33, which was destroyed during its collision with the dead Russian satellite Cosmos 2251 on February 10, 2009, was launched by a Russian Proton K rocket from the Russian-leased Tyuratam (Baikonur Cosmodrome) facility in Kazakhstan. On March 4, 1998, Russia informed the United Nations that "[O]n 14 September 1997, seven Iridium satellites were placed in Earth orbit by a single Proton carrier rocket from the Baikonur launch site. . . . The satellites are owned and operated by the Motorola Company (of the United States)."[29] In fact these satellites were actually owned by Iridium, an international industrial consortium of which Motorola was the leading shareholder.

On the other hand, the official U. S. Registry of Space Objects Launched into Outer Space, maintained by the U. S. Department of State's Bureau of Oceans and International Environmental and Scientific Affairs (with jurisdiction over the Iridium 33, for which the United States is the flag state) affirms that the satellite was not registered with the United Nations by the United States, and that it was "[M]entioned by Russian Federation in ST/SG/SER.E/332."[30] Moreover, as of May 22, 2013, the U. S. Registry still showed that the Iridium 33 satellite was in orbit, though it had been destroyed more than 4 years earlier.

Japan, though it does not license its own (or its universities') small satellites that have been launched by foreign launch vehicles, does registers them nationally. Japan registered its SDS-4 (Small Demonstration Satellite 4), a small satellite weighing about 50 kg and measuring 50 cm with the United Nations.[31] This small satellite was developed and launched by JAXA on May 17, 2012, from Tanegashima Space Center, Kagoshima, Japan.

Nigeriasat-II and Nigeriasat-X have not yet been registered with the United Nations, but it is believed that the international registration process will certainly be followed in due course. Nigeria had submitted the required registration information to the United Nations for registering Nigeriasat-I (the precursor of Nigeriast II) and that satellite was registered in 2005 pursuant to U. N. General Assembly Resolution 1721 B (XVI) because, at that time, Nigeria had not acceded to the Registration Convention.[32]

---

[29] Information Furnished In Conformity With The Convention On Registration Of Objects Launched Into Outer Space. Note verbale dated 4 March 1998 from the Permanent Mission of the Russian Federation to the United Nations addressed to the Secretary-General. UN Doc. ST/SG/SER.E/332 of 19 March 1998, Vienna, United Nations.

[30] U.S. Space Objects Registry, 2 Nov. 2009; available online at: http://usspaceobjectsregistry.state.gov/registry/dsp_DetailView.cfm?id=1517&searched=1.

[31] Information furnished in conformity with the Convention on Registration of Objects Launched into Outer Space, Note verbale dated 12 October 2012 from the Permanent Mission of Japan to the United Nations (Vienna) addressed to the Secretary-General, UN Doc. ST/SG/SER.E/655 of 18 October 2012.

[32] Information furnished in conformity with General Assembly resolution 1721 B (XVI) by States launching objects into orbit or beyond: Note verbale dated 17 August 2004 from the Permanent Mission of Nigeria to the United Nations (Vienna) addressed to the Secretary-General, UN Doc. A/AC.105/INF.411 of 31 March 2005.

It is expected that with an exponential increase in the launch of small satellites, especially pico satellites or nano satellites or cube-sats that will remain in orbit for a very short period of time, states may not be registering them internationally with the U. N. Secretary-General and perhaps not even on their own national registries. That will eventually cause problems in their identification, particularly if they happen to be involved in accidents in outer space (as was the case of Ecuador's Pegaso cube-sat), or if they survive on re-entry and cause damage on Earth.

5.3   Use of Radio Frequencies

For their proper functioning, all satellites – big and small – need to use radio frequencies whether they are telecommunications, remote sensing, scientific spacecraft, or spacecraft designed for other types of applications. In order to avoid possible harmful interference, radio frequencies are heavily regulated both at the international and national levels. Radio frequencies are a limited international natural resource to be used by all countries on an equitable basis, and they do not respect national borders.[33] Therefore, the international community has devised an extensive international regulatory system through the International Telecommunication Union (ITU), which is the oldest specialized agency of the United Nations. The regulatory control of interference is achieved mainly through ITU (international) radio regulations, which are international treaties applicable to 193 ITU member states. These regulatory mechanisms relating to control of interference are in the form of: (a) allocation of separate radio frequencies to different radio services; (b) power limits on the transmission of radio signals by all stations; (c) coordination of use of radio frequencies among various states (administrations within the ITU); and (d) requirement of protection of the stations in the geostationary satellite orbit (GSO) by non-GSO satellites.[34] All these international regulatory requirements are also applicable to small and micro satellites. Despite all of these protections, the level and incidence of unintended and intentional radio interference continues to increase. Organizations such as the Satellite Interference Reduction Group (sIRG) have been formed to seek avenues to mitigate this problem.

Many of the small satellites use radio frequencies that are allocated to the Amateur Satellite Service under the ITU radio regulations. A significant number of other "small satellites" that are deployed in constellations for telecommunications or machine-to-machine data relay use frequencies allocated to communications services, while yet others use allocations for scientific or other purposes. The Amateur Satellite Service is defined as a radio communication service using space stations on

---

[33] Article 44 (2), ITU Constitution.

[34] Attila Matas, "ITU Radio Regulations Related to Small Satellites", a presentation made at 10th Annual Cube-sat Developers' Workshop 2013, 24–26 April 2013, Cal Poly, San Luis Obispo, USA; available online at: http://mstl.atl.calpoly.edu/~bklofas/Presentations/DevelopersWorkshop 2013/Matas_ITU_Regulations_for_Small_Satellites.pdf.

Earth for the purpose of self-training, intercommunication, and technical investigations carried out by amateurs, that is, by duly authorized persons interested in radio technique solely with a personal aim and without pecuniary interest.[35] In the Table of Frequency Allocations under Article 5 of the ITU radio regulations, specific bands of radio frequency allocations are included for amateur satellite service on the same basis for three ITU regions. Several technical and regulatory limitations are placed on the use of these bands. Within only a few years of the commencement of the launching of small satellites, Amateur Satellite Service bands have been getting increasingly crowded mainly due to the increasing number of such satellites.[36] Concerns are being expressed about this situation, which could be expected to get worse when more small satellites are launched.

In order to address the problem of overcrowding in Amateur Satellite Service bands by small satellites, some governments have submitted various proposals to the ITU's World Radiocommunication Conference that was held in Geneva in 2012 (WRC-12). The conference in its Resolution COM6/10 (WRC-12) recognized that (a) currently, many nano and pico satellites use spectrum allocated to the Amateur Satellite Service and the MetSat service in the frequency range 30–3,000 MHz, although their missions are potentially inconsistent with these services, and (b) the existing provisions of the ITU radio regulations for coordination and notification of satellites under Articles 9 and 11 may need to be adapted to take account of the nature of these satellites. The WRC-12 invited the ITU-Radiocommunications Bureau to examine the procedures for notifying space networks and to consider possible modifications to enable the deployment and operation of nano and pico satellites, taking into account the short development time, short mission time, and unique orbital characteristics. WRC-12 also instructed the director of the Radiocommunication Bureau to report to WRC-15 on the results of these studies. The conference also invited WRC-18 to consider whether modifications to the regulatory procedures for notifying satellite networks are needed to facilitate the deployment and operation of nano and pico satellites, and to take the appropriate actions. This is a positive action that might help in the accommodation of the growing need for sufficient radio frequencies by small and micro satellites.

Under the ITU Constitution, Convention, and Radio Regulations, each state is obliged to fulfill the following requirements:

- Allow its small satellite operators only to use radio frequencies as allocated under the ITU radio regulations.
- Prevent them from causing harmful interference to the radio services of others,
- Require them to operate their satellites in accordance with the ITU radio regulations.
- Require them to obtain licenses from the designated governmental agency.

---

[35] Article 1 (paras 1.56 and 1.57), ITU Radio Regulations.

[36] "Basic Space Technology Initiative (BSTI): Activities in 2011–2012 and plans for 2013 and beyond," UN Doc. A/AC.105/2012/CRP.16 of 23 May 2012, p. 4.

There is, however, no enforcement provisions or sanctions that the ITU can invoke to ensure compliance with these "requirements."

Since most of the small satellites use radio frequencies allocated to the Amateur Satellite Service, the concerned administrations (states or countries) are expected to respect the following requirements specified in the ITU radio regulations:

- "No. 25.11 – Administrations authorizing space stations in the Amateur Satellite Service shall ensure that sufficient Earth command stations are established before launch to ensure that any harmful interference caused by emissions from a station in the amateur-satellite service can be terminated immediately (see No. 22.1).
- No. 22.1 – Space stations shall be fitted with devices to ensure immediate cessation of their radio emissions by tele-command, whenever such cessation is required under the provisions of these regulations."

In certain cases, the concerned country (administration) of a small satellite may be further obliged to notify to, coordinate through, and register with ITU if in the use of radio frequencies of transmitting and receiving Earth and space stations: (1) international protection against harmful interference is desired, (2) the radio frequencies will be used for international service, or (3) it is believed that the use of a new radio frequencies will cause harmful interference to other stations. National implementation of, and compliance with, the ITU requirements and procedures could be a highly cumbersome, time consuming, and expensive process for the operators of small satellites.

In the United States, all satellites are required to be licensed by the Federal Communications Commission (FCC) under the Communications Act of 1934 (as amended) and the Federal Satellite Communications Regulations.[37] In addition, the FCC also imposes, in some cases, international regulations and procedures as specified in ITU radio regulations. Therefore, the FCC procedures for a radio license could be a lengthy and time-consuming process for small satellites, especially when they need to be coordinated internationally through the ITU.

The FCC recognizes the difficulties that are faced by small satellites and has thus recently issued simplified guidelines to provide guidance to small satellite operators concerning FCC licensing for the use of radio frequencies allocated to Amateur Satellite Service.[38] This certainly will be helpful to the operators of small satellites (particularly, pico satellites, nano satellites, and femto satellites) since most of them are generally unaware of the complexities of the FCC satellite-licensing regime. The same also applies with regard to parallel requirements in other countries.

---

[37] Communications Act of 1934, (47 USC Chapter 5); available online at: http://www.law.cornell.edu/uscode/text/47/chapter-5; and Satellite Communications Regulations, (47 Code of Federal Regulations Part 25); available online at: http://www.law.cornell.edu/cfr/text/47/25.

[38] Federal Communications Commission, *GUIDANCE ON OBTAINING LICENSES FOR SMALL SATELLITES*, Public Notice Released: March 15, 2013.

Chapter 6
Responsibility, Liability, and Orbital Debris Mitigation Issues

As noted in Chap. 1 the amount of space debris in Earth orbit has been increasing, particularly as a result of the intentional destruction of space objects and recent accidents. The most significant causes of new debris generation in recent years has been the collision of the Russian Cosmos 2251 and the Iridium 33 satellite, the missile destruction of a defunct Chinese meteorological satellite, and, most recently, the encounter between an Ecuadorian small satellite and space debris.

It is believed that the safe and sustainable use of outer space will increasingly continue to be threatened by space debris, firstly as a navigation hazard to operational satellites due to collisions in space and secondly as a major risk to humans, property, and the environment on the surface of Earth. An example of the latter risk is the re-entry of Russian satellite Cosmos 954 in 1978 that scattered radioactive debris over a large area of northern Canada.

The expected launch of small satellites, particularly in low Earth orbits, will further expand the amount of space debris, as these satellites generally have short life spans. This is of particular concern because historically, nano-satellites have a relatively high failure rate of 52 % even though new technology and quality small satellite kits that are now available are expected to increase reliability.

Through several efforts both at international and national levels, some regulatory provisions, technical standards and voluntary guidelines have been adopted to regulate and/or to mitigate the generation of space debris.

6.1 Responsibility and Liability Issues

As noted earlier, states are internationally responsible for ensuring that the space activities of their own and private citizens (including the launching and operation of small satellites) are carried out in conformity with the 1967 Outer Space Treaty. Such activities of non-governmental entities must be carried out pursuant to "authorization and continuing supervision" by states, and, more importantly, in the conduct of their space activities, due regard must be paid to the corresponding interests of all other

R.S. Jakhu and J.N. Pelton, *Small Satellites and Their Regulation*, SpringerBriefs in Space Development, DOI 10.1007/978-1-4614-9423-2_6, © Springer New York 2014

states as well as their public and private entities. A state may be obliged to make reparations for damage caused as a result of its contravention of these obligations.

The increasing growth of small satellites and the potential for enhanced space debris have given rise to several regulatory issues, the most important of which relates to the responsibility and liability for damage caused by a space object, including even damage caused by small satellites. The issue of responsibility and liability has been addressed under the currently applicable international space law. Below, we provide a brief description of the applicability or otherwise of international space law dealing with liability.

Due to privatization and expansion of space activities and other reasons, it is possible that many small satellites may not even end up being governed by international space law. Thus, in terms of general international law, the question remains unanswered as to what the responsibility and liability regime is when it comes to space objects deployed by entities other than nations and in those states that are not parties to space treaties such as the 1967 Outer Space Treaty and the 1972 Liability Convention. Another important question relates to national regulation of such satellites, particularly in those countries that do not have appropriate legal regulations in place.

6.1.1 Liability for Small Satellites as Space Objects

Small satellites are like any other space objects involving the launching of payloads, the possibility of collisions in outer space, and the likelihood of debris falling back to Earth from the failed or dead system(s). However, in some ways, constellations of small satellites will have different implications from many other space objects launched and operated to date. Even if there are significant advances in launch technologies, the launch activities for small satellites will be enormous and thus may be associated with higher probabilities of mishaps causing damage on orbit in outer space, in the air, and/or on the surface of Earth.

There are two main international space law treaties that directly apply to the cases of liability for damage occurring during the conduct of space activities, including the launching and operation of small satellites; i.e., the 1967 Outer Space Treaty and the 1972 Liability Convention. A state party to these treaties, or its citizens, has the option to make a claim for compensation under either of these agreements if damage suffered is caused by any other state party to these two agreements. It is important to keep in mind that compensation for any damage caused by a space object or its component parts, a launch vehicle or its component parts, or any piece of debris created by them, will be recoverable under either or both of these treaties.

The Outer Space Treaty, under its Article VII, holds a launching state liable if the damage is caused "to another state party to the Treaty or to its natural or juridical persons by such object or its component parts on the Earth, in air space or in outer space." A "launching state" is the state that "launches or procures the launching of an object into outer space, including the moon and other celestial bodies, and each state party from whose territory or facility an object is launched." In addition, under Article VI of the Outer Space Treaty, each state party to the treaty is internationally responsible for national activities in outer space, whether such activities are carried on by governmental agencies or by non-governmental entities. Space activities carried out by non-governmental entities shall require authorization and continuing supervision by the appropriate state. A state is also responsible for such activities if carried out by an international organization in which that state participates.

It is only in the English version of the treaty that a distinction is made between responsibility and liability. However, the texts of the treaty in the Chinese, French, Russian, and Spanish languages, which are equally authentic, make no distinction between responsibility and liability. Therefore, a liability claim for compensation can also be made under Article VI of the Outer Space Treaty. The treaty neither places limitations on the amount of liability nor defines the term 'damage.' Therefore, the ordinary meaning of the term (i.e., loss of or harm to one's property or injury to or death of a natural person) will likely be used.

The amount of compensation to be claimed could be such as would be sufficient to restore the injured party to, where possible, the situation that existed before the damage occurred. This could include not only direct damages but also indirect, mental, moral, and consequential damages. The amount of compensation is to be determined in accordance with international law and the principles of justice and equity. However, if the case is brought before a national court, the court would generally apply its national law to make that determination.

The provisions of Article VII of the Outer Space Treaty have been further elaborated upon and strengthened by the Liability Convention. The Convention, in unambiguous terms, holds the launching state "absolutely liable to pay compensation for damage caused by its space object on the surface of the Earth or to aircraft in flight."[1] This feature of the convention made possible the straightforward and expedient settlement of Canada's claim against the U.S.S.R. for damage caused by a dead Soviet space object (i.e., space debris) Cosmos 954 when it intruded into Canadian air space, depositing on Canadian territory hazardous radioactive debris that was carried onboard.

On the other hand, Article III of the Liability Convention creates fault-based liability for damage caused in outer space "to a space object of one launching state or to persons or property onboard such a space object by a space object of another launching state." In other words, the claimant state must establish not only that the damage has been caused by a space object (or its component parts or debris created by it) belonging to another state, but also that the damage was due to the latter state's fault or the fault of persons for whom that state was responsible. It should be noted that due to limited space monitoring (space surveillance) capability, especially on the part of a claimant state that is not a developed space power, it will be difficult, if not impossible, to clearly and convincingly establish fault on the part of the state whose small satellite (including an untracked small piece of space debris) is believed to have caused the damage. In such cases, perhaps the expertise of a third state or its private company that possesses appropriate monitoring capability may be used to provide the required data about a particular accident. The court that is

---

[1] Article II, the Liability Convention.

settling the dispute involving damage caused by a small satellite may also use witnesses who have expertise in space surveillance.

According to Article I (a) of the Liability Convention, the term "damage" "means loss of life, personal injury, or other impairment of health; or loss of or damage to property of states or of persons, natural or juridical, or property of international intergovernmental organizations." Some authors believe that only physical damage caused by a small satellite (or its component parts or debris) would be recoverable. However, since compensation for "other impairment of health" is recoverable, it is reasonable to assume that mental or psychological damage without any physical manifestation would also be covered by the term 'damage.' International space law only imposes liability for damage on states and not upon their private entities. There are and will certainly be numerous private companies, academic institutions, and even NGO's undertaking the construction and operation of small satellite systems.

If a private company or academic institution builds and/or launches a small satellite, its state of nationality (being a launching state) should have a provision under its domestic law for licensing to facilitate that state's performance of its international obligation of 'authorization and continuing supervision' as required under Article VI of the Outer Space Treaty. It will be difficult for states, especially those that do not have adequate or appropriate national space laws, to regulate the activities of such entities. Thus they themselves will end up bearing the total cost of compensation, if required to pay, to third parties. The authorizing state would not only be responsible but could also be held liable if any damage is caused by a small satellite owned by the authorized company or academic institution of that state. Similarly, states are responsible and could be held liable if a small satellite causes any damage even where that small satellite was not specifically authorized. Being highly visible and possibly rich, the state that launches (or whose private company launches) a small satellite could become an easy target for lawsuits for compensation, especially when the satellite that caused damage belongs to a small country, small company, or an academic institution.

6.1.2   Liability Under General International Law or National Law

A state that is not a party to the Outer Space Treaty or the Liability Convention or its citizens who suffer any damage caused by the space activity of a foreign state may hold that foreign state liable if there is a genuine link between the foreign state and its entity whose satellite caused the damage, and make a claim of state responsibility under general international law or under the national law of the state that is believed to have caused the damage.[2] However, generally, such claims would have to overcome severe uncertainty (such as choice of law, conflicts of law, grounds for claims, recoverability and quantum of damages, court procedure, nature and

---

[2] *Draft Articles on Responsibility of States for Internationally Wrongful Acts*, adopted by the International Law Commission at its fifty-third session (2001), from the Report of the International Law Commission on the work of its Fifty-third session, *Official Records of the General Assembly, Fifty-sixth session, Supplement No. 10* (A/56/10), (chp.IV.E.1).

admissibility of evidence, language of the court, and jurisdiction of the court). Such efforts may also be extremely expensive and perhaps could drag on for a fairly long period of time before being resolved.

6.1.3 Risk Management

States or their entities that exposure themselves to liability for damage caused by their small satellites could manage the risk of liability by procuring insurance coverage. Insurance can be obtained by the satellite owner, launch suppliers, or the satellite operator. In some countries, obtaining liability insurance is a condition for procurement of a launch and/operation license. For example in the United States before a launch license can be issued the applicant must secure sufficient liability insurance or demonstrate financial responsibility in amounts to compensate for the maximum probable loss that may arise from damage claims against the U. S. government for third party death, bodily injury, or property damage. However, it is possible that small entities, academic institutions, or small countries may not be aware of the applicable international or national regulatory requirements, the potential liability risks, or possible insurance coverage or may even consider the cost of insurance to be more expensive than that of the satellite itself. It will be in the interest of launching states to have appropriate regulatory regime(s) in place that prescribe mandatory insurance requirements for the launch of small satellites.

6.2 Orbital Debris Mitigation Issues

International space law (particularly, the 1967 Outer Space Treaty and the 1972 Liability Convention) does not specifically address the regulation of the problem of the generation of space debris, though it does deal with the consequences if damage is caused by a piece of space debris. On the other hand, there has been a trend against the negotiation of any international treaty and a preference for non-binding guidelines. The issue of space debris has been discussed for about 20 years within COPUOS[3] and outside the United Nations.

The following three sets of space debris mitigation standards and guidelines have been adopted:

- The European Space Debris Safety and Mitigation Standard issued by the European Space Agency (ESA) (2002)
- The Inter-Agency space Debris Coordination Committee (IADC) Space Debris Mitigation Guidelines (2002)
- The U. N. Committee on the Peaceful Uses of Outer Space (COPUOS) Space Debris Mitigation Guidelines (2007)

---

[3] "Twentieth Year of Space Debris Discussions at the United Nations", NASA, Orbital Debris: Quarterly News, April 2013, p. 1.

These guidelines, which are internationally observed on a voluntary basis without any enforcement sanctions, in brief, specify the following practices:

- Limit debris released during normal operations;
- Minimize potential for break-ups during operational phases;
- Limit the probability of accidental collision in orbit;
- Avoid intentional destruction and other harmful activities;
- Minimize potential for post-mission break-ups resulting from stored energy;
- Limit the long-term presence (up to 25 years) of spacecraft and launch vehicle orbital stages in low Earth orbit after the end of their mission; and
- Limit the long-term interference of spacecraft and launch vehicle orbital stages with geosynchronous region after the end of their mission.

It is important to note that these "voluntary guidelines" are not legally binding under international law. States and international organizations are expected to voluntarily take measures, through national mechanisms or through their own applicable mechanisms, to ensure that these guidelines are implemented to the greatest extent feasible through space debris mitigation practices and procedures. Most importantly, these guidelines do not deal with remediation of existing space debris, which has immense potential to create new debris as a result of fragmentation.

Domestic application of these guidelines, particularly to small and micro satellites, will be challenging, especially in those countries that do not have national regulatory mechanisms, or do not issue or require any license for such satellites.

Although registered in Japan, Japanese small satellites have traditionally been launched by foreign launch vehicles and the foreign countries concerned usually address their licensing, safety and debris mitigation requirements. The safety of such satellites and debris mitigation measures associated therewith will only be ensured when a JAXA rocket launches them. In that case, JAXA's safety regulations and debris standards will be imposed.

In order to reduce space debris created by satellites, some technical solutions are being researched and explored, for example, the design and development of a sail system or micro-thrusters that will pull small satellites out of orbit to avoid increasing the amount of space debris.[4] It has been suggested that instead of being a problem, small satellites could be used to remove pieces of dead satellites, thereby helping to solve the space debris problem.[5] However, the practical viability of this and other proposed technical solutions and regulatory mechanisms have not as yet been fully explored.

---

[4] Stephen Harris, "Cube-sat sail system is able to pull small satellites out of orbit," 23 November 2012; available online at http://www.theengineer.co.uk/aerospace/news/cube-sat-sail-system-is-able-to-pull-small-satellites-out-of-orbit/1014720.article; Jennifer Chu, "MIT-developed 'micro-thrusters' could propel small satellites," Aug 17, 2012; available online at http://phys.org/news/2012-08-mit-developed-microthrusters-propel-small-satellites.html.

[5] Jose Guerrero, et al., "How can Small Satellites be used to Support Orbital Debris Removal Goals Instead of increasing the problem?", a paper presented at the 24th Annual AIAA/USU Conference on Small Satellites, 2010; available online at: http://digitalcommons.usu.edu/cgi/viewcontent.cgi?article=1197&context=smallsat.

Chapter 7
Technical, Operational, and Regulatory Solutions to Small Satellite Issues

There are a number of challenges facing those who would like to design, build, launch, and operate small satellite projects. These can be generally divided into the following categories: (i) technical and operational; and (ii) legal, regulatory, and liability concerns. Although these issues and concerns are interrelated we will address them in this order and from an interdisciplinary perspective.

7.1 Technical and Operational Issues

In Chap. 3, we noted the many advances that are being made with regard to small satellites. New technologies are being developed to make small satellites more capable, reliable, and better able to de-orbit at the end of life. Also, new launch capabilities are being developed to facilitate the cost-effective deployment of small satellites into desired orbits. Standardization of systems and development of quality kits can serve to increase reliability while reducing costs. In addition to the foregoing, much more can be done to develop new technology that can aid the cause of low-cost and reliable small satellites. This applies to almost every aspect of a small satellite's development.

7.1.1 Challenges for New Power Systems

In the area of power, improved lithium ion and lithium-carbon ion batteries represent the highest energy density currently available, but there is the issue of these types of batteries bursting into flames. Research into this area is clearly needed. In the area of solar cells, quantum dot technology and multiple junction photovoltaic cells are promising areas of research. In the case of the lowest cost small satellites, silicon-based solar cells can continue to be effectively used. In this case, the objective may be to develop and use lower cost solar cells using either amorphous or symmetrical silicon-based wafers to provide the lowest cost source of solar energy.

R.S. Jakhu and J.N. Pelton, *Small Satellites and Their Regulation*, SpringerBriefs in Space Development, DOI 10.1007/978-1-4614-9423-2_7, © Springer New York 2014

7.1.2   Challenges for New Antenna Systems

In the area of antennas, the goal is to find the lowest cost and reliable antenna systems that can still provide higher gain performance. In this regard, inflatable antennas may be a cost-effective line of research and development. This concept has a further appeal in that the inflating mechanism for the high gain antenna could also be used to inflate balloons or other mechanisms to help with more efficient de-orbit at the end of mission life. To the extent that lower-cost phased-array feed systems could be developed to work with inflatable (or other low cost and low mass antenna reflectors), this could serve to create a higher performance and yet lower-cost antenna system.

7.1.3   Challenges for New De-orbiting Systems

Considerable progress has been made in recent years to develop improved systems to facilitate the de-orbit of small satellites. Many of these involve inflatable systems that are some form of balloon or inflatable structure that greatly increases the cross-section of the small satellite and thus expands the atmospheric drag that is experienced. There could be variations on such a theme. One might be able to combine an inflatable antenna with an extendable boom for gravity gradient stabilization and thus combine a functional system capability with a de-orbit mechanism. Such a system would be deployed at the start of a mission and thus reduce the overall lifetime of the project. Such a design might still allow useful operation for 5–7 years. Another concept that might be explored is the deployment of several (three or four) small satellites at the same time with linked lightweight-booms. At the end of life, a thin Mylar film might be extended to create a significant drag across the frame that connects the free-flying small satellites. The main point at this stage is to define an ongoing development goal to find new, low cost and more effective means to de-orbit small satellites. Recent progress in this area at different research sites and academic institutions is encouraging and demonstrates that solutions can indeed be found.

7.1.4   Challenges for New Positioning and Pointing Systems

Recent projects have increasingly shown that active thrusters can indeed be developed to improve positioning and pointing systems. Water and alcohol thrusters have been demonstrated to be reasonable in cost and reliability and are reasonably effective. The 4-unit to 8-unit thruster systems seem to be quite viable and promising. Although these positioning and pointing systems are designed to support payload mission objectives, they could also provide some assistance at end of life.

7.1.5 Challenges for Standardization and Kit Systems

The movement toward dimension standardization a 1-unit, a 2-unit, and up to an 8-unit cube-sat and the development of design-optimized kits has clearly extended the range of potential small satellite experimenters by "de-complicating" the process of "building" a small satellite. It may now be the time to move toward a new generation of advanced "consolidator kits" that could allow a number of cube-sat units to be "plugged into" a larger unit that would combine a number of experiments or even application packages together not only for consolidated launch and deployment but also for a deployable de-orbit system at the end of life. Since the consolidated system with de-orbit capabilities would be of value to all future users of space as a global commons, it might be possible to create a fund that would provide economic incentives to encourage the use of these consolidated "mega cube" systems and/or to assist with launch and deployment. In some ways, this approach would be the moral equivalent of Nanorack experiments that would fly on the space station, but it would still allow for those who wish to have "free flying" satellites, but in such a way as to minimize debris and with streamlined registration procedures.

In addition there are other technological challenges to design kits that are more capable, lighter in mass, or present greater capabilities. Clearly, these improvements should be pursued as well.

7.2 Regulatory, Legal, and Liability Issues

Technological advancements and operational innovations cannot resolve all of the issues and provide solutions for all of the concerns that relate to the future of small satellites. Innovations on the regulatory front can also be helpful. It is often thought that rules and regulations and new legal conditions will only add cost, complexity and extended schedules to small satellite projects. In some cases this is true. Yet in the United States, the new procedures related to small satellites actually streamline registration processes for small satellite projects. Certainly, new regulatory processes related to consolidating small satellite projects are aimed at creating positive incentives for student exploration and college-based experimental projects, among others. Consolidating free-flyer small satellites into integrated activities will not only save launch and operational costs but also streamline registration and regulatory requirements.

All states that actually launch satellites should, before the launch, enter into an agreement that requires from the owner of the satellites (a) proper end-of-life disposal of satellites, (b) compulsory insurance against third party claims, and (c) a wide range of protective and mitigating action as outlined in Sect. 7.1.

Regulatory changes are necessary to reduce the regulatory burden and to expand the availability of sufficient radio frequencies, to micro satellites. In this regard, the efforts that are being made by the ITU as well as the FCC and NOAA are good and should be followed by other states. Yet, there need to be corresponding steps to address the related potential problem of creating more space debris.

International registration of all satellites ought to be followed more effectively. One may think of electronically linking national registries to the international registry with the United Nations. Increased possibility of liability for damage by small satellites could impose undue regulatory and financial challenges upon their owners and operators. Thus, some sort of balance would be needed in the form of appropriate technical and regulatory measures. These measures might include international monitoring capabilities, space situational awareness, space traffic coordination, bilateral agreements between launch provider and satellite operator; insurance requirements, and debris mitigation and removal measures.

Small satellites can represent new opportunities for organizations and nations with limited financing to conduct space experiments. Their small size and mass can greatly reduce launch costs. However, they present a potential serious space safety dilemma in terms of orbital debris. There are risks that can be minimized in a number of ways, as discussed earlier in terms of adding active or passive systems to aid de-orbit, combining efforts into larger systems with active controls, or carrying out experiments via missions that involve the use of the International Space Station, hosted payloads, or even private space stations wherein the experiment is taken up and down and is never a free flyer.

There are a number of ethical and technical guidelines that apply to satellites that have come into focus in recent years. At present, none of these guidelines particularly distinguish small satellites from large ones. No nano satellite project could exist if all of the guidelines that apply to large satellites were rigorously imposed in the case of small satellites. Launch providers and associated liability partners bear a part of the burden as to what practical and ethical guidelines they follow in accepting the launch of small satellite missions. At least, they might become more discriminating of the small satellites they agree to deploy and the missions of the satellites.

Clearly, the small satellite community must pay greater attention to operational and design best practices. The International Standards Organization (ISO) has begun developing such non-normative operational practices. The case of operational constellations involving the use of small satellites in the range of 1,000 kg or more involves a different set of issues and ethical considerations. In these cases, active de-orbit must be considered essential, and designs such as the Teledesic "megasatellite" constellation should be considered unacceptable until and unless totally new technologies related to collision avoidance and de-orbit is somehow developed in the future.

Chapter 8
Ten Top Things to Know About Small Satellites and Space Debris

Because of the substantial advantages that small satellites can offer to educational programs on limited budgets, to those deploying global constellations in low Earth orbit, and to those who can meet specific space-related tasks with miniaturized payloads, it seems likely that there will continue to be an exponential increase in launch of small satellites and consequently, in the number of associated debris in orbit. In order to reduce space debris, national regulatory and technical solutions would need to be developed and implemented. In light of the fear of a future avalanche of space debris that can cascade out of control, there are many steps that are being taken to mitigate space debris and a number of these steps prescribe better ways to proceed with small satellite missions and experiments.

The following represent our top ten thoughts about how the small satellite enterprise might move forward in a positive way without making the already serious space debris problem even worse.

1. **There are many new and promising technologies that should be supported with targeted R&D to aid better small satellite design and operation. Consolidated "mega cube" satellite systems and kits that include de-orbit capabilities would be of value and could streamline registration procedures.**

   Enormous progress has been made in the past decade to develop amazing new technologies. The new technologies related to micro thrusters, high speed processing, end-of-life de-orbit systems, on-board storage, and standardized kits have combined to allow better small satellite design, lower cost, higher performance at lower cost with higher reliability and longer life. Yet there is much more that can still be done to develop better miniaturized components (i.e., microprocessors and storage units), improved power systems (i.e., quantum dot solar power systems, improved batteries), lower cost and more compact de-orbit systems, improved antenna systems (i.e., phased-array feed systems, inflatable antennas), and lower cost thrusters and launch systems. In addition to the development of new satellite and launch technologies, there can also be improved technology transfer systems that allow developments related to larger-scale spacecraft to be applied to smaller-scale systems.

R.S. Jakhu and J.N. Pelton, *Small Satellites and Their Regulation*, SpringerBriefs in Space Development, DOI 10.1007/978-1-4614-9423-2_8, © Springer New York 2014

2. De-orbit and pointing and positioning systems for small satellites should be considered a priority. Again consolidated free-flyers might be a cost-effective way to accomplish this goal.

One of the large differences between early types of small satellites and many that are being designed and deployed today is that the latter can be equipped with active thrusters for positioning and de-orbiting capabilities. There are many new technologies that might be considered useful for small satellite missions, but the development of new mechanisms that provide reliable and low cost positioning capabilities and especially systems that can aid removal from orbit at end-of life should be considered the top priority in terms of new development programs. These could be a combination of "active" programs, such as small scale and low level thrusters, or "passive" in terms of inflatable balloons or wings that increase atmospheric drag and assist eventual removal from orbit. Elements such as inflatable antennas could add to small satellite capabilities in the first phase of operation and then assist with removal from orbit at end-of-life.

3. "Consolidation" for many types of small satellite projects or examination of a "hosted payload" approach to meeting mission needs should be considered a prime objective and implemented whenever possible.

Many small-scale space projects should be examined at the earliest stages to see whether multiple mission objectives could be achieved through "effective consolidation." Often, such consolidation can reduce costs and risks and also minimize problems associated with orbital debris. There are many options now available in this respect. One increasingly attractive alternative is the use of hosted payloads. This approach can be used for one-of-a-kind experiments where a package can be hosted on a geosynchronous satellite. In other instances where global coverage is required, various types of packages might be hosted on low Earth orbit constellations when numerous subsystem packages need to fly.

Yet another option is for small educational space projects to use the Nanoracks capability that permits experiments to fly on the International Space Station. In this case the costs are low and astronauts can start and stop experiments while also providing dynamic control over them. Further, by running the experiments on the International Space Station there is no problem of de-orbiting nano satellites at the end-of-life. As private space stations are deployed such as the Bigelow Aerospace Company intends to do, the range of options for flying a wide variety of experiments and educational packages that are small, medium, or large will multiply. The idea that small, nano, pico, or femto satellites must be free flyers in space really has no particular advantage other than some sort of assumed "national or personal prestige." Combining and consolidating small satellite missions as "packages" that can share power and fly into space as an integrated effort has many advantages. This consolidated approach can cut launch, satellite mission-design, and operational costs. It can also extend experimental times in orbit, reduce potential liabilities, insure access to reliable power, aid capabilities in such areas as pointing accuracy, positioning, and stability. Finally it could ease the difficulty of meeting registration requirements.

4. **All spacefaring nations and enterprises that launch satellites should agree to binding arrangements for orbital debris mitigation and active debris removal.**

   Significant progress has been made through the IADC collaborations and the U.N.'s COPUOS to move to agreement on voluntary procedures to prevent the creation of new orbital debris and to remove objects at end-of life. These procedures, however, need to be strengthened and made mandatory in nature. They need to be transformed into binding international law backed by sanctions (and or rewards) to help enforce them. Today, many small satellites are not being registered and they are thus, in a way, "flying under the radar" when it comes to careful monitoring and concerted efforts to avoid their possible contribution to the orbital debris problem.

   The problem of orbital debris is, of course, not just the result of satellite deployment but also debris that is created by upper stage launch vehicles and such things as exploding fuel tanks. All types of activities that contribute to orbital debris need to be considered and mandatory procedures developed to mitigate this problem. Many that are new to space activities or have limited resources or wish to use small satellites for experiments might tend to feel that they are being discriminated against because they have not created the problem of space debris, but they are being singled out for some of the most restrictive measures. In this regard, larger spacefaring nations with assets in space that allow small scale space experiments (i.e. the owners and operators of the International Space Station) might wish to give consideration to incentives such as permitting educational experiments and projects from countries new to space use to be consolidated on experimental facilities like Nanoracks. In so doing, large spacefaring nations stand to benefit in the long run by avoiding the proliferation of free-flyers that contribute to the space debris problem.

5. **New economic arrangements and insurance provisions should be put into place for all satellite launches, including small satellites.**

   As discussed above, new regulations may not only forestall the creation of new debris but could also require each new launch contribute to a fund to support active debris removal. Today, as most commercial launches into orbit take place, there is a launch insurance policy in place that provides various types of coverage. Some of the coverage is for liability in the event that a major accident should occur and the mission should fly off course and land in a populated area. Some of the coverage is for the mission itself and offers financial protection in the event that the objectives of the mission fail to be achieved. This coverage would serve to pay for a new launch and a new satellite. There are other types of insurance coverage to protect against a satellite collision and the destruction of the satellite.

   There is no reason that insurance mechanisms and a related orbital debris fund could not be fashioned to cope with the problem of space debris. This new type of economic arrangement and insurance coverage would apply to all launches (commercial, civilian, government, and defense) and all types and classes of satellites – large, medium, and small. The need for insurance (particularly with regard to liability and protection against orbital debris) could be, in effect, eliminated for these individual efforts if small space missions were carried out as consolidated projects launched as combined packages. A consolidated mission would also simplify registration notifications since there could be only one registration rather than four if that many projects were indeed consolidated This type of insurance arrangement would thus act as both a "carrot" and a "stick" (i.e., reward and/or sanction) in that the cost of a small "free flyer" would be more, but the cost of a "consolidated" mission would be reduced.

6. **New liability arrangements for space objects as well as incentives for removal of space debris need to be put in place.**

   The current Liability Convention is not well suited to fully addressing orbital debris issues. Currently, a nation only pays for damage from a space object if it occurs and liability is clearly established. There is no particular reward or incentive to actively work to prevent debris from occurring and to minimize the risk in the first place. A positive step would be to amend the Liability Convention so that countries and commercial organizations have active incentives to reduce space debris in the first place and have legal and economic processes that would help to minimize risk and reduce future potential liability from the outset.

   Unfortunately, this does not seem likely to occur in the relative near term. Thus, it might be necessary for the members of the IADC to discuss this issue and see if there might be some sort of formula that spacefaring nations might agree to – including the creation of a multi-lateral space object liability fund that would cover a first round of liability claims in the event of a space debris-related accident. Such an arrangement might serve as a basis for addressing this issue and establish a possible transfer of liability exposure from one space actor to another with the fund serving as possible form of "insurer" against an undesirable outcome. It would seem that some such mechanism could reduce the overall risk and create incentives to reduce the risk of space collisions as well as the creation of more space debris in the future. In short, we need space agreements and mechanisms to allow solutions to be pursued actively, rather than just hoping that space accidents do not occur. If these types of economic arrangements cannot be devised, then perhaps mandatory arbitration procedures could be set up as yet another way to approach this problem.

   If such liability reforms cannot be done through the mechanism of COPUOS and other U.N. processes, then perhaps other options may be possible. New types of liability insurance arrangements and new ways to pay to actively reduce orbit debris risks might be discussed and agreed through the IADC, or perhaps even more likely through the space insurance business. What seems necessary is some new types of governmentally sanctioned or commercially agreed arrangements. These arrangements need to be backed up by sanctions, financial bonds, or some form of insurance or mandatory arbitration arrangements.

7. **De-orbit provisions should be lowered from 25 years to 20 years and be made mandatory.**

   Currently, the IADC guidelines recommend that satellites be designed so that they will de-orbit within 25 years. With the increasing rate of launch of cube-sats, nano satellites, pico satellites, and even femto satellites many view these provisions as no longer being adequate. The addition of inflatables or other de-orbit mechanisms should make such new guidelines feasible without undue complexity and expense. To the extent that addition of de-orbit capabilities is seen as an undue burden, the possibility of moving to consolidated experimentation on the International Space Station (i.e., via Nanoracks) or on commercial space stations such as Bigelow Aerospace is planning to deploy can offer longer cost-saving options. The change, however, needs to include mandatory registration of small satellites and de-orbit provisions that are backed by some form of reward or sanction process. This would entail agreement by all spacefaring nations to not launch any small satellite unless suitable arrangements are made for de-orbit within 20 years or to migrate the small satellite mission to a consolidated mission on a space station or a hosted payload so that the return of the small satellite to Earth would be implemented on a guaranteed basis.

8. **Part of the longer term solution of orbit debris and space safety would seem to require some form of space traffic management and control that is achieved through the International Civil Aviation Organization (ICAO) and/or national and regional air and space traffic agencies.**

The systematic study of space traffic management and control is now in its earliest stages. Preliminary steps have included the publishing of books on this issue such as *The Need for an Integrated Regulatory Regime for Aviation and Space: An ICAO for Space?* Edited by Ram S. Jakhu, Tommaso Sgobba, and Paul S. Dempsey. Discussions are now underway to create a study process within the International Association for the Advancement of Space Safety (IAASS) in cooperation with ICAO and national or regional air traffic control agencies.

The initial focus of these new processes will give top priority to the safety of airline passengers as the most significant risk factor to mitigate. As the number of private space activities such as suborbital space adventure flights, commercial launches to space, private space stations, and especially hypersonic transportation tests increases, the range of issues to be explored and new regulatory capabilities to be devised will also increase. As these regulatory efforts increase over time it is important for the range of issues to also expand to cover environmental, frequency management, as well as other issues and concerns that arise from space transportation and flights that involve so-called sub-space or the "protozone" operations in altitudes that range between 21 and 100 km – the normally accepted definition of outer space.[1] These activities may eventually involve the amendment of the 1944 Convention on International Civil Aviation (Chicago Convention), under which ICAO operates, and formal designation of responsibilities to U.N., agencies such as the World Meteorological Organization (WMO), the U.N. Environmental Program (UNEP), and the International Telecommunication Union (ITU).

---

[1] Joseph N. Pelton, "Beyond the Protozone: A New Global Regulatory Regime for Air and Space" American Bar Association Forum on Air and Space Law, Washington, D.C. June 6, 2013.

9. **New international regulations and guidelines should be put into place with regard to toxic rocket and thruster fuels and power systems, etc.**

Again, these concerns are not specifically related to small satellites since issues involving toxic rocket and thruster fuels and hazardous power systems relate first and foremost to larger and medium-sized satellite deployment. As more environmental friendly fuels and power systems are developed, they will need to be applied to all types of satellites – including small satellites. Since owners and operators of small satellite missions are extremely cost conscious, there will be a particular concern to make sure that new regulations in these areas, as well as restrictions related to de-orbiting systems, positioning, etc., do not create undue financial difficulties or create overly difficult regulatory processes for those engaged in small satellite-related activities. Currently, there are particularly difficult issues to be addressed in that some of the safest and lowest cost rocket systems that have been developed by commercial launch systems involve the burning of solid fuel (i.e., neoprene) which is particularly challenging in terms of damaging particulate emissions and environmental concerns.

10. **It is imperative to undertake active debris removal activities pursuant to an international operational and regulatory framework that should establish an inter-governmental organization incorporating public-private partnerships.**[2]

There is a massive amount of debris already in existence in Earth orbit that now exceeds 6,300 metric tons. In order to avoid the generation of new debris that results in a catastrophic type Kessler syndrome, active debris removal of *existing* debris seems more and more essential in addition to the mitigation and prevention efforts. Various technical means and debris removal capabilities are being developed. However, the removal of space objects faces numerous challenges, both technical and regulatory.

The state on whose registry an object launched into outer space is carried holds jurisdiction and control over even a non-functional space object (i.e., space debris). If a state, or a state-licensed actor, wishes to remove a space object, it can only legally do so if it has legal jurisdiction and control over that space object (i.e., space debris) or with prior permission from the state of registry. Regulatory mechanisms must be sought to facilitate the seeking and granting of permission and to establish rules respecting both the jurisdiction and control issue and consent. There should be a standard and legally acceptable definition of what constitutes space debris in order to permit the conduct of active debris removal activities. Moreover, active debris removal technologies and activities have strategic and military implications since they may be used as anti-satellite weapons (ASAT). In order to minimize military, diplomatic, and

---

[2] For details, see "Active Debris Removal – An Essential Mechanism for Ensuring the Safety and Sustainability of Outer Space: A Report of the International Interdisciplinary Congress on Space Debris Remediation and On-Orbit Satellite Servicing," UN Document: A/AC.105/C.1/2012/CRP.16 of 27 January 2012.

political concerns in relation to debris removal or changing the orbit of any space object, it is believed that debris removal activities need to be monitored and coordinated at both the national and international levels, and should be undertaken pursuant to an international operational and regulatory framework. This might be accomplished via commercial arrangements or perhaps even require the establishment of an inter-governmental organization (IGO) to foster the development of the technologies for active debris removal and subsequently to perform, "license," or coordinate the removal operations on a commercial basis. The international agreement establishing such an organization should have (a) a clear definition of space debris, and (b) a provision under which the participating states authorize the removal or servicing of those pieces of space debris for which they are the states of registration. All these considerations are complicated by the fact that there are a host of other space-related issues that also need to be addressed, such as space traffic management, environmental protection, and regulation of the "protozone." New international arrangements in these areas may or may not overlap with arrangements involving space debris.

The bottom line is that space activities will become more and more of a political, economic, legal, ethical, and commercial interest to all people and nations. It is time – indeed it is past the time – for comprehensive thought and action to be given to the best ways for dealing with such problems. We hope that the United Nations Committee on the Peaceful Uses of Outer Space (COPUOS) and the Inter-Agency Space Debris Coordinating (IADC) Committee, along with other relevant international agencies, will start to seriously address space-related issues that now face us and seek new solutions before they become even more difficult to solve.